PRAISE FOR *WAKING UP*

"An extraordinary and ambitious masterwork. . . . altogether spectacular."

—Maria Popova, *Brain Pickings*

"Harris's book . . . caught my eye because it's so entirely of this moment, so keenly in touch with the growing number of Americans who are willing to say that they do not find the succor they crave, or a truth that makes sense to them, in organized religion."

—Frank Bruni, columnist, *The New York Times*

"The fact is that *Waking Up* lends a different picture of Harris (at least to me): an intelligent and sensitive person who is willing to undergo the discomfort involved in proposing alternatives to the religions he's spent years degrading. His new book, whether discussing the poverty of spiritual language, the neurophysiology of consciousness, psychedelic experience, or the quandaries of the self, at the very least acknowledges the potency and importance of the religious impulse—though Harris might name it differently— that fundamental and common instinct to seek not just an answer to life, but a way to live that answer."

—Trevor Quirk, *The New Republic*

"Uber-atheist Sam Harris is getting all spiritual. In his new book, *Waking Up: A Guide to Spirituality Without Religion*, the usually outspoken critic of religion describes how spirituality can and must be divorced from religion if the human mind is to reach its full potential. . . . But there is plenty in *Waking Up* that will delight Harris's most militant atheist readers."

—*Religion News Service*

"The great value and novelty of this book is that Harris, in a simple but rigorous style, takes the middle way between these pseudoscientific and pseudospiritual assertions . . . [leading] to a profoundly more salubrious life."

—*Publishers Weekly*

"A demanding, illusion-shattering book."

—*Kirkus Reviews*

"Don't read *Waking Up* . . . if you want to be told that heaven is real. Do read it if you want to explore the nature of consciousness, to learn how just trying to be mindful can free you from anxiety and self-blame."

—*MORE Magazine*

"*Waking Up* is an eye-opening, mind-expanding book."

—AA Agnostica

"A seeker's memoir, a scientific and philosophical exploration of the self, and a how-to guide for transcendence, *Waking Up* explores the nature of consciousness, explains how to meditate, tells you the best drugs to take, and warns you about lecherous gurus. It will shake up your most fundamental beliefs about everyday experience, and it just might change your life."

—Paul Bloom, professor of psychology and cognitive science, Yale University, and author of *Just Babies: The Origins of Good and Evil*

"*Waking Up* is a rigorous, kind, clear, and witty book that will point you toward the selflessness that is our original nature."

—Stephen Mitchell

"Sam Harris points out the rational methodology for exploring the nature of consciousness and for experiencing a transformative understanding of possibilities. *Waking Up* really does help us wake up."

—Joseph Goldstein, author of *Mindfulness:
A Practical Guide to Awakening* and *One Dharma*

"As a neuroscientist, Sam Harris shows how our egos are illusions, diffuse products of brain activity; and as a long-term practitioner of meditation, he shows how abandoning this illusion can wake us up to a richer life, more connected to everything around us."

—Jerry Coyne, professor of biology at the University of Chicago and author of *Why Evolution is True*

"Sam Harris ranks as my favorite skeptic, bar none. In *Waking Up* he gives us a clear-headed, no-holds-barred look at the spiritual supermarket, calling out what amounts to junk food and showing us where real nutrition can be found. Anyone who realizes the value of a spiritual life will find much to savor here—and those who see no value in it will find much to reflect on."

—Daniel Goleman, author *Emotional Intelligence* and *Focus*

"Sam Harris has written a beautifully rational book about spirituality, consciousness, and transcendence. He is the high priest of spirituality without religion. I recommend this book regardless of your belief system. As befits a book called *Waking Up*, it's an eye opener."

—A. J. Jacobs, bestselling author of *The Year of Living Biblically*

ALSO BY SAM HARRIS

The End of Faith

Letter to a Christian Nation

The Moral Landscape

Free Will

Lying

Waking Up

A GUIDE TO SPIRITUALITY WITHOUT RELIGION

SAM HARRIS

SIMON & SCHUSTER PAPERBACKS

New York London Toronto Sydney New Delhi

For Annaka, Emma, and Violet

Simon & Schuster Paperbacks
An Imprint of Simon & Schuster, Inc.
1230 Avenue of the Americas
New York, NY 10020

First Simon & Schuster trade paperback edition June 2015

SIMON & SCHUSTER PAPERBACKS and colophon are
registered trademarks of Simon & Schuster, Inc.

For information about special discounts for bulk purchases, please contact
Simon & Schuster Special Sales at 1-866-506-1949 or
business@simonandschuster.com.

The Simon & Schuster Speakers Bureau can bring authors to your live event.
For more information or to book an event contact the
Simon & Schuster Speakers Bureau at 1-866-248-3049 or
visit our website at www.simonspeakers.com.

Book design by Ellen R. Sasahara

Manufactured in the United States of America

15 17 19 20 18 16 14

The Library of Congress has cataloged the hardcover edition as follows:

Harris, Sam
Waking up : a guide to spirituality without religion / Sam Harris.—First Simon
& Schuster hardcover edition.
pages cm
1. Spirituality. 2. Irreligion. I. Title.
BL624.H333 2014
204—dc23   2014012097

ISBN 978-1-4516-3601-7
ISBN 978-1-4516-3602-4 (pbk)
ISBN 978-1-4516-3603-1 (ebook)

Some portions of this book appeared previously on the author's website.

CONTENTS

Waking Up

Chapter 1

Spirituality

I once participated in a twenty-three-day wilderness program in the mountains of Colorado. If the purpose of this course was to expose students to dangerous lightning and half the world's mosquitoes, it was fulfilled on the first day. What was in essence a forced march through hundreds of miles of backcountry culminated in a ritual known as "the solo," where we were finally permitted to rest—alone, on the outskirts of a gorgeous alpine lake—for three days of fasting and contemplation.

I had just turned sixteen, and this was my first taste of true solitude since exiting my mother's womb. It proved a sufficient provocation. After a long nap and a glance at the icy waters of the lake, the promising young man I imagined myself to be was quickly cut down by loneliness and boredom. I filled the pages of my journal not with the insights of a budding naturalist, philosopher, or mystic but with a list of the foods on which I intended to gorge myself the instant I returned to civilization. Judging from the state of my consciousness at the time, millions of years of hominid evolution had produced nothing more transcendent than a craving for a cheeseburger and a chocolate milkshake.

ce of sitting undisturbed for three days
d starlight, with nothing to do but con-
my existence, to be a source of perfect
ld see not so much as a glimmer of my
ters home, in their plaintiveness and self-
t Shiloh or Gallipoli.

So I was more than a little surprised when several members of our party, most of whom were a decade older than I, described their days and nights of solitude in positive, even transformational terms. I simply didn't know what to make of their claims to happiness. How could someone's happiness *increase* when all the material sources of pleasure and distraction had been removed? At that age, the nature of my own mind did not interest me—only my life did. And I was utterly oblivious to how different life would be if the quality of my mind were to change.

Our minds are all we have. They are all we have ever had. And they are all we can offer others. This might not be obvious, especially when there are aspects of your life that seem in need of improvement—when your goals are unrealized, or you are struggling to find a career, or you have relationships that need repairing. But it's the truth. Every experience you have ever had has been shaped by your mind. Every relationship is as good or as bad as it is because of the minds involved. If you are perpetually angry, depressed, confused, and unloving, or your attention is elsewhere, it won't matter how successful you become or who is in your life—you won't enjoy any of it.

Most of us could easily compile a list of goals we want to achieve or personal problems that need to be solved. But what is the real significance of every item on such a list? Everything we want to accomplish—to paint the house, learn a new language, find a better job—is something that promises that, if done, it would allow us to finally relax and enjoy our lives in the present. Generally speaking, this is a false hope. I'm not denying the importance of achieving one's goals, maintaining one's health, or keeping one's children clothed and fed—but most of us spend our time seeking happiness and security without acknowledging the underlying purpose of our search. Each of us is looking for a path back to the present: We are trying to find good enough reasons to be satisfied *now*.

Acknowledging that this is the structure of the game we are playing allows us to play it differently. How we pay attention to the present moment largely determines the character of our experience and, therefore, the quality of our lives. Mystics and contemplatives have made this claim for ages—but a growing body of scientific research now bears it out.

A few years after my first painful encounter with solitude, in the winter of 1987, I took the drug 3,4-methylenedioxy-N-methylamphetamine (MDMA), commonly known as Ecstasy, and my sense of the human mind's potential shifted profoundly. Although MDMA would become ubiquitous at dance clubs and "raves" in the 1990s, at that time I didn't know anyone of my generation who had tried it. One evening, a few months before my twentieth birthday, a close friend and I decided to take the drug.

The setting of our experiment bore little resemblance to the conditions of Dionysian abandon under which MDMA is now often consumed. We were alone in a house, seated across from each other on opposite ends of a couch, and engaged in quiet conversation as the chemical worked its way into our heads. Unlike other drugs with which we were by then familiar (marijuana and alcohol), MDMA produced no feeling of distortion in our senses. Our minds seemed completely clear.

In the midst of this ordinariness, however, I was suddenly struck by the knowledge that I loved my friend. This shouldn't have surprised me—he was, after all, one of my best friends. However, at that age I was not in the habit of dwelling on how much I loved the men in my life. Now I could *feel* that I loved him, and this feeling had ethical implications that suddenly seemed as profound as they now sound pedestrian on the page: *I wanted him to be happy.*

That conviction came crashing down with such force that something seemed to give way inside me. In fact, the insight appeared to restructure my mind. My capacity for envy, for instance—the sense of being diminished by the happiness or success of another person—seemed like a symptom of mental illness that had vanished without a trace. I could no more have felt envy at that moment than I could have wanted to poke out my own eyes. What did I care if my friend was better looking or a better athlete than I was? If I could have bestowed those gifts on him, I would have. *Truly* wanting him to be happy made his happiness my own.

A certain euphoria was creeping into these reflections, perhaps, but the general feeling remained one of absolute sobriety—and of moral and emotional clarity unlike any I had ever known. It would not be too strong to say that I felt sane for the first time in my life. And yet the change in my consciousness seemed entirely straightforward. I was simply talking to my friend—about what, I don't recall—and realized that I had ceased to be concerned about myself. I was no longer anxious, self-critical, guarded by irony, in competition, avoiding embarrassment, ruminating about the past and future, or making any other gesture of thought or attention that separated me from him. I was no longer watching myself through another person's eyes.

And then came the insight that irrevocably transformed my sense of how good human life could be. I was feeling *boundless* love for one of my best friends, and I suddenly realized that if a stranger had walked through the door at that moment, he or she would have been fully included in this love. Love was at bottom impersonal—and deeper than any personal history could justify. Indeed, a transactional form of love—I love you *because* . . . —now made no sense at all.

The interesting thing about this final shift in perspective was that it was not driven by any change in the way I felt. I was not overwhelmed by a new feeling of love. The insight had more the character of a geometric proof: It was as if, having glimpsed the properties of one set of parallel lines, I suddenly understood what must be common to them all.

The moment I could find a voice with which to speak, I discovered that this epiphany about the universality of love could be readily communicated. My friend got the point at once: All I had to do was ask him how he would feel in the presence of a total stranger at that moment, and the same door opened in his mind. It was simply obvious that love, compassion, and joy in the joy of others extended without limit. The experience was not of love growing but of its being no longer obscured. Love was—as advertised by mystics and crackpots through the ages—a state of being. How had we not seen this before? And how could we overlook it ever again?

It would take me many years to put this experience into context. Until that moment, I had viewed organized religion as merely a monument to the ignorance and superstition of our ancestors. But I now knew that Jesus, the Buddha, Lao Tzu, and the other saints and sages of history had not all been epileptics, schizophrenics, or frauds. I still considered the world's religions to be mere intellectual ruins, maintained at enormous economic and social cost, but I now understood that important psychological truths could be found in the rubble.

———

Twenty percent of Americans describe themselves as "spiritual but not religious." Although the claim seems to annoy believers and atheists equally, separating spirituality from religion is a perfectly reasonable thing to do. It is to assert two important truths simultaneously: Our world is dangerously riven by religious doctrines that all educated people should condemn, and yet there is more to understanding the human condition than science and secular culture generally admit. One purpose of this book is to give both these convictions intellectual and empirical support.

Before going any further, I should address the animosity that many readers feel toward the term *spiritual*. Whenever I use the word, as in referring to meditation as a "spiritual practice," I hear from fellow skeptics and atheists who think that I have committed a grievous error.

The word *spirit* comes from the Latin *spiritus*, which is a translation of the Greek *pneuma*, meaning "breath." Around the thirteenth century, the term became entangled with beliefs about immaterial souls, supernatural beings, ghosts, and so forth. It acquired other meanings as well: We speak of the *spirit* of a thing as its most essential principle or of certain volatile substances and liquors as *spirits*. Nevertheless, many nonbelievers now consider all things "spiritual" to be contaminated by medieval superstition.

I do not share their semantic concerns.[1] Yes, to walk the aisles of any "spiritual" bookstore is to confront the yearning and credulity of our species by the yard, but there is no other term—apart from the even more problematic *mystical* or the more restrictive *contemplative*—with which to discuss the efforts people make, through meditation, psychedelics, or other means, to fully bring their minds into the present or to induce nonordinary states of consciousness. And no other word links this spectrum of experience to our ethical lives.

Throughout this book, I discuss certain classically spiritual phenomena, concepts, and practices in the context of our modern understanding of the human mind—and I cannot do this while restricting myself to the terminology of ordinary experience. So I will use *spiritual*, *mystical*, *contemplative*, and *transcendent* without further apology. However, I will be precise in describing the experiences and methods that merit these terms.

For many years, I have been a vocal critic of religion, and I won't ride the same hobbyhorse here. I hope that I have been sufficiently energetic on this front that even my most skeptical readers will trust that my bullshit detector remains well calibrated as we advance over this new terrain. Perhaps the following assurance can suffice for the moment: Nothing in this book needs to be accepted on faith. Although my focus is on human subjectivity—I am, after all, talking about the nature of experience itself—all my assertions can be tested in the laboratory of your own life. In fact, my goal is to encourage you to do just that.

Authors who attempt to build a bridge between science and spirituality tend to make one of two mistakes: Scientists generally start with an impoverished view of spiritual experience, assuming that it must be a grandiose way of describing ordinary states of mind—parental love, artistic inspiration, awe at the beauty of the night sky. In this vein, one finds Einstein's amazement at the intelligibility of Nature's laws described as though it were a kind of mystical insight.

New Age thinkers usually enter the ditch on the other side of the road: They idealize altered states of consciousness and draw specious connections between subjective experience and the spookier theories at the frontiers of physics. Here we are told that the Buddha and other contemplatives anticipated modern cosmology or quantum mechanics and that by transcending the sense of self, a person can realize his identity with the One Mind that gave birth to the cosmos.

In the end, we are left to choose between pseudo-spirituality and pseudo-science.

Few scientists and philosophers have developed strong skills of introspection—in fact, most doubt that such abilities even exist. Conversely, many of the greatest contemplatives know nothing about science. But there is a connection between scientific fact and spiritual wisdom, and it is more direct than most people suppose. Although the insights we can have in meditation tell us nothing about the origins of the universe, they do confirm some well-established truths about the human mind: Our conventional sense of self is an illusion; positive emotions, such as compassion and patience, are teachable skills; and the way we think directly influences our experience of the world.

There is now a large literature on the psychological benefits of meditation. Different techniques produce long-lasting changes in attention, emotion, cognition, and pain perception, and these correlate with both structural and functional changes in the brain. This field of research is quickly growing, as is our understanding of self-awareness and related mental phenomena. Given recent advances in neuroimaging technology, we no longer face a practical impediment to investigating spiritual insights in the context of science.

Spirituality *must* be distinguished from religion—because people of every faith, and of none, have had the same sorts of spiritual experiences. While these states of mind are usually interpreted through the lens of one or another religious doctrine, we know that this is a mistake. Nothing that a Christian, a Muslim, and a Hindu can experience—self-transcending love, ecstasy, bliss, inner light—constitutes evidence in support of their traditional beliefs, because their beliefs are logically incompatible with one another. A deeper principle must be at work.

That principle is the subject of this book: The feeling that we call "I" is an illusion. There is no discrete self or ego living like a Minotaur in the labyrinth of the brain. And the feeling that there is—the sense of being perched somewhere behind your eyes, looking out at a world that is separate from yourself—can be altered or entirely extinguished. Although such experiences of "self-transcendence" are generally thought about in religious terms, there is nothing, in principle, irrational about them. From both a scientific and a philosophical point of view, they represent a clearer understanding of the way things are. Deepening that understanding, and repeatedly cutting through the illusion of the self, is what is meant by "spirituality" in the context of this book.

Confusion and suffering may be our birthright, but wisdom and happiness are available. The landscape of human experience includes deeply transformative insights about the nature of one's own consciousness, and yet it is obvious that these psychological states must be understood in the context of neuroscience, psychology, and related fields.

I am often asked what will replace organized religion. The answer, I believe, is nothing and everything. Nothing need replace its ludicrous and divisive doctrines—such as the idea that Jesus will return to earth and hurl unbelievers into a lake of fire, or that death in defense of Islam is the highest good. These are terrifying and debasing fictions. But what about love, compassion, moral goodness, and self-transcendence? Many people still imagine that religion is the true repository of these virtues. To change this, we must talk about the full range of human experience in a way that is as free of dogma as the best science already is.

This book is by turns a seeker's memoir, an introduction to the brain, a manual of contemplative instruction, and a philosophical unraveling of what most people consider to be the center of their inner lives: the feeling of self we call "I." I have not set out to describe all the traditional approaches to spirituality and to weigh their strengths and weaknesses. Rather, my goal is to pluck the diamond from the dunghill of esoteric religion. There is a diamond there, and I have devoted a fair amount of my life to contemplating it, but getting it in hand requires that we remain true to the deepest principles of scientific skepticism and make no obeisance to tradition. Where I do discuss specific teachings, such as those of Buddhism or Advaita Vedanta, it isn't my purpose to provide anything like a comprehensive account. Readers who are loyal to any one spiritual tradition or who specialize in the academic study of religion, may view my approach as the quintessence of arrogance. I consider it, rather, a symptom of impatience. There is barely time enough in a book—or in a life—to get to the point. Just as a modern treatise on weaponry would omit the casting of spells and would very likely ignore the slingshot and the boomerang, I will focus on what I consider the most promising lines of spiritual inquiry.

My hope is that my personal experience will help readers to see the nature of their own minds in a new light. A rational approach to spirituality seems to be what is missing from secularism and from the lives of most of the people I meet. The purpose of this book is to offer readers a clear view of the problem, along with some tools to help them solve it for themselves.

THE SEARCH FOR HAPPINESS

*One day, you will find yourself outside this world which is like
a mother's womb. You will leave this earth to enter, while you
are yet in the body, a vast expanse, and know that the words,
"God's earth is vast," name this region from which the saints
have come.*

Jalal-ud-Din Rumi

I share the concern, expressed by many atheists, that the terms *spiritual* and *mystical* are often used to make claims not merely about the quality of certain experiences but about reality at large. Far too often, these words are invoked in support of religious beliefs that are morally and intellectually grotesque. Consequently, many of my fellow atheists consider all talk of spirituality to be a sign of mental illness, conscious imposture, or self-deception. This is a problem, because millions of people have had experiences for which *spiritual* and *mystical* seem the only terms available. Many of the beliefs people form on the basis of these experiences are false. But the fact that most atheists will view a statement like Rumi's above as a symptom of the man's derangement grants a kernel of truth to the rantings of even our least rational opponents. The human mind does, in fact, contain vast expanses that few of us ever discover.

And there *is* something degraded and degrading about many of our habits of attention as we shop, gossip, argue, and ruminate our way to the grave. Perhaps I should speak only for myself here: It seems to me that I spend much of my waking life in a neurotic trance. My experiences in meditation suggest, however, that an alternative exists. It is possible to stand free of the juggernaut of self, if only for moments at a time.

Most cultures have produced men and women who have found that certain deliberate uses of attention—meditation, yoga, prayer—can transform their perception of the world. Their efforts generally begin with the realization that even in the best of circumstances, happiness is elusive. We seek pleasant sights, sounds, tastes, sensations, and moods. We satisfy our intellectual curiosity. We surround ourselves with friends and loved ones. We become connoisseurs of art, music, or food. But our pleasures are, by their very nature, fleeting. If we enjoy some great professional success, our feelings of accomplishment remain vivid and intoxicating for an hour, or perhaps a day, but then they subside. And the search goes on. The effort required to keep boredom and other unpleasantness at bay must continue, moment to moment.

Ceaseless change is an unreliable basis for lasting fulfillment. Realizing this, many people begin to wonder whether a deeper source of well-being exists. Is there a form of happiness beyond the mere repetition of pleasure and avoidance of pain? Is there a happiness that does not depend upon having one's favorite foods available, or friends and loved ones within arm's reach, or good books to read, or something to look forward to on the weekend? Is it possible to be happy *before* anything happens, before one's desires are gratified, in spite of life's difficulties, in the very midst of physical pain, old age, disease, and death?

We are all, in some sense, living our answer to this question—and most of us are living as though the answer were "no." No, nothing is more profound than repeating one's pleasures and avoiding one's pains; nothing is more profound than seeking satisfaction—sensory, emotional, and intellectual—moment after moment. Just keep your foot on the gas until you run out of road.

Certain people, however, come to suspect that human existence might encompass more than this. Many of them are led to suspect this by *religion*—by the claims of the Buddha or Jesus or some other celebrated figure. And such people often begin to practice various disciplines of attention as a means of examining their experience closely enough to see whether a deeper source of well-being exists. They may even sequester themselves in caves or monasteries for months or years at a time to facilitate this process. Why would a person do this? No doubt there are many motives for retreating from the world, and some of them are psychologically unhealthy. In its wisest form, however, the exercise amounts to a very simple experiment. Here is its logic: If there exists a source of psychological well-being that does not depend upon merely gratifying one's desires, then it should be present even when all the usual sources of pleasure have been removed. Such happiness should be available to a person who has declined to marry her high school sweetheart, renounced her career and material possessions, and gone off to a cave or some other spot that is inhospitable to ordinary aspirations.

One clue to how daunting most people would find such a project is the fact that solitary confinement—which is essentially what we are talking about—is considered a punishment *inside* a maximum-security prison. Even when forced to live among murderers and rapists, most people still prefer the company of others to spending any significant amount of time alone in a room. And yet contemplatives in many traditions claim to experience extraordinary depths of psychological well-being while living in isolation for vast stretches of time. How should we interpret this? Either the contemplative literature is a catalogue of religious delusion, psychopathology, and deliberate fraud, or people have been having liberating insights under the name of "spirituality" and "mysticism" for millennia.

Unlike many atheists, I have spent much of my life seeking experiences of the kind that gave rise to the world's religions. Despite the painful results of my first few days alone in the mountains of Colorado, I later studied with a wide range of monks, lamas, yogis, and other contemplatives, some of whom had lived for decades in seclusion doing nothing but meditating. In the process, I spent two years on silent retreat myself (in increments of one week to three months), practicing various techniques of meditation for twelve to eighteen hours a day.

I can attest that when one goes into silence and meditates for weeks or months at a time, doing nothing else—not speaking, reading, or writing, just making a moment-to-moment effort to observe the contents of consciousness—one has experiences that are generally unavailable to people who have not undertaken a similar practice. I believe that such states of mind have a lot to say about the nature of consciousness and the possibilities of human well-being. Leaving aside the metaphysics, mythology, and sectarian dogma, what contemplatives throughout history have discovered is that there is an alternative to being continuously spellbound by the conversation we are having with ourselves; there is an alternative to simply identifying with the next thought that pops into consciousness. And glimpsing this alternative dispels the conventional illusion of the self.

Most traditions of spirituality also suggest a connection between self-transcendence and living ethically. Not all good feelings have an ethical valence, and pathological forms of ecstasy surely exist. I have no doubt, for instance, that many suicide bombers feel extraordinarily good just before they detonate themselves in a crowd. But there are also forms of mental pleasure that are intrinsically ethical. As I indicated earlier, for some states of consciousness, a phrase like "boundless love" does not seem overblown. It is decidedly inconvenient for the forces of reason and secularism that if someone wakes up tomorrow feeling boundless love for all sentient beings, the only people likely to acknowledge the legitimacy of his experience will be representatives of one or another Iron Age religion or New Age cult.

Most of us are far wiser than we may appear to be. We know how to keep our relationships in order, to use our time well, to improve our health, to lose weight, to learn valuable skills, and to solve many other riddles of existence. But following even the straight and open path to happiness is hard. If your best friend were to ask how she could live a better life, you would probably find many useful things to say, and yet you might not live that way yourself. On one level, wisdom is nothing more profound than an ability to follow one's own advice. However, there are deeper insights to be had about the nature of our minds. Unfortunately, they have been discussed entirely in the context of religion and, therefore, have been shrouded in fallacy and superstition for all of human history.

The problem of finding happiness in this world arrives with our first breath—and our needs and desires seem to multiply by the hour. To spend any time in the presence of a young child is to witness a mind ceaselessly buffeted by joy and sorrow. As we grow older, our laughter and tears become less gratuitous, perhaps, but the same process of change continues: One roiling complex of thought and emotion is followed by the next, like waves in the ocean.

Seeking, finding, maintaining, and safeguarding our well-being is the great project to which we all are devoted, whether or not we choose to think in these terms. This is not to say that we want mere pleasure or the easiest possible life. Many things require extraordinary effort to accomplish, and some of us learn to enjoy the struggle. Any athlete knows that certain kinds of pain can be exquisitely pleasurable. The burn of lifting weights, for instance, would be excruciating if it were a symptom of terminal illness. But because it is associated with health and fitness, most people find it enjoyable. Here we see that cognition and emotion are not separate. The way we think about experience can completely determine how we feel about it.

And we always face tensions and trade-offs. In some moments we crave excitement and in others rest. We might love the taste of wine and chocolate, but rarely for breakfast. Whatever the context, our minds are perpetually moving—generally toward pleasure (or its imagined source) and away from pain. I am not the first person to have noticed this.

Our struggle to navigate the space of possible pains and pleasures produces most of human culture. Medical science attempts to prolong our health and to reduce the suffering associated with illness, aging, and death. All forms of media cater to our thirst for information and entertainment. Political and economic institutions seek to ensure our peaceful collaboration with one another—and the police or the military is summoned when they fail. Beyond ensuring our survival, civilization is a vast machine invented by the human mind to regulate its states. We are ever in the process of creating and repairing a world that our minds want to be in. And wherever we look, we see the evidence of our successes and our failures. Unfortunately, failure enjoys a natural advantage. Wrong answers to any problem outnumber right ones by a wide margin, and it seems that it will always be easier to break things than to fix them.

Despite the beauty of our world and the scope of human accomplishment, it is hard not to worry that the forces of chaos will triumph—not merely in the end but in every moment. Our pleasures, however refined or easily acquired, are by their very nature fleeting. They begin to subside the instant they arise, only to be replaced by fresh desires or feelings of discomfort. You can't get enough of your favorite meal until, in the next moment, you find you are so stuffed as to nearly require the attention of a surgeon—and yet, by some quirk of physics, you still have room for dessert. The pleasure of dessert lasts a few seconds, and then the lingering taste in your mouth must be banished by a drink of water. The warmth of the sun feels wonderful on your skin, but soon it becomes too much of a good thing. A move to the shade brings immediate relief, but after a minute or two, the breeze is just a little too cold. Do you have a sweater in the car? Let's take a look. Yes, there it is. You're warm now, but you notice that your sweater has seen better days. Does it make you look carefree or disheveled? Perhaps it is time to go shopping for something new. And so it goes.

We seem to do little more than lurch between wanting and not wanting. Thus, the question naturally arises: Is there more to life than this? Might it be possible to feel much better (in *every sense* of *better*) than one tends to feel? Is it possible to find lasting fulfillment despite the inevitability of change?

Spiritual life begins with a suspicion that the answer to such questions could well be "yes." And a true spiritual practitioner is someone who has discovered that it is possible to be at ease in the world for no reason, if only for a few moments at a time, and that such ease is synonymous with transcending the apparent boundaries of the self. Those who have never tasted such peace of mind might view these assertions as highly suspect. Nevertheless, it is a fact that a condition of selfless well-being is there to be glimpsed in each moment. Of course, I'm not claiming to have experienced all such states, but I meet many people who appear to have experienced none of them—and these people often profess to have no interest in spiritual life.

This is not surprising. The phenomenon of self-transcendence is generally sought and interpreted in a religious context, and it is precisely the sort of experience that tends to increase a person's faith. How many Christians, having once felt their hearts grow as wide as the world, will decide to ditch Christianity and proclaim their atheism? Not many, I suspect. How many people who have never felt anything of the kind become atheists? I don't know, but there is little doubt that these mental states act as a kind of filter: The faithful count them in support of ancient dogma, and their absence gives nonbelievers further reason to reject religion.

This is a difficult problem for me to address in the context of a book, because many readers will have no idea what I'm talking about when I describe certain spiritual experiences and might assume that the assertions I'm making must be accepted on faith. Religious readers present a different challenge: They may think they know exactly what I'm describing, but only insofar as it aligns with one or another religious doctrine. It seems to me that both these attitudes present impressive obstacles to understanding spirituality in the way that I intend. I can only hope that, whatever your background, you will approach the exercises presented in this book with an open mind.

RELIGION, EAST AND WEST

We are often encouraged to believe that all religions are the same: All teach the same ethical principles; all urge their followers to contemplate the same divine reality; all are equally wise, compassionate, and true within their sphere—or equally divisive and false, depending on one's view.

No serious adherents of any faith can believe these things, because most religions make claims about reality that are mutually incompatible. Exceptions to this rule exist, but they provide little relief from what is essentially a zero-sum contest of all against all. The polytheism of Hinduism allows it to digest parts of many other faiths: If Christians insist that Jesus Christ is the son of God, for instance, Hindus can make him yet another avatar of Vishnu without losing any sleep. But this spirit of inclusiveness points in one direction only, and even it has its limits. Hindus are committed to specific metaphysical ideas—the law of karma and rebirth, a multiplicity of gods—that almost every other major religion decries. It is impossible for any faith, no matter how elastic, to fully honor the truth claims of another.

Devout Jews, Christians, and Muslims believe that theirs is the one true and complete revelation—because that is what their holy books say of themselves. Only secularists and New Age dabblers can mistake the modern tactic of "interfaith dialogue" for an underlying unity of all religions.

I have long argued that confusion about the unity of religions is an artifact of language. *Religion* is a term like *sports*: Some sports are peaceful but spectacularly dangerous ("free solo" rock climbing); some are safer but synonymous with violence (mixed martial arts); and some entail little more risk of injury than standing in the shower (bowling). To speak of sports as a generic activity makes it impossible to discuss what athletes actually do or the physical attributes required to do it. What do all sports have in common apart from breathing? Not much. The term *religion* is hardly more useful.

The same could be said of *spirituality*. The esoteric doctrines found within every religious tradition are not all derived from the same insights. Nor are they equally empirical, logical, parsimonious, or wise. They don't always point to the same underlying reality—and when they do, they don't do it equally well. Nor are all these teachings equally suited for export beyond the cultures that first conceived them.

Making distinctions of this kind, however, is deeply unfashionable in intellectual circles. In my experience, people do not want to hear that Islam supports violence in a way that Jainism doesn't, or that Buddhism offers a truly sophisticated, empirical approach to understanding the human mind, whereas Christianity presents an almost perfect impediment to such understanding. In many circles, to make invidious comparisons of this kind is to stand convicted of bigotry.

In one sense, all religions and spiritual practices must address the same reality—because people of all faiths have glimpsed many of the same truths. Any view of consciousness and the cosmos that is available to the human mind can, in principle, be appreciated by anyone. It is not surprising, therefore, that individual Jews, Christians, Muslims, and Buddhists have given voice to some of the same insights and intuitions. This merely indicates that human cognition and emotion run deeper than religion. (But we knew that, didn't we?) It does not suggest that all religions understand our spiritual possibilities equally well.

One way of missing this point is to declare that all spiritual teachings are inflections of the same "Perennial Philosophy." The writer Aldous Huxley brought this idea into prominence by publishing an anthology by that title. Here is how he justified the idea:

> *Philosophia perennis*—the phrase was coined by Leibniz; but the thing—the metaphysic that recognizes a divine Reality substantial to the world of things and lives and minds; the psychology that finds in the soul something

similar to, or even identical with, divine Reality; the ethic that places man's final end in the knowledge of the immanent and transcendent Ground of all being—the thing is immemorial and universal. Rudiments of the Perennial Philosophy may be found among the traditionary lore of primitive peoples in every region of the world, and in its fully developed forms it has a place in every one of the higher religions. A version of this Highest Common Factor in all preceding and subsequent theologies was first committed to writing more than twenty-five centuries ago, and since that time the inexhaustible theme has been treated again and again, from the standpoint of every religious tradition and in all the principal languages of Asia and Europe.[2]

Although Huxley was being reasonably cautious in his wording, this notion of a "highest common factor" uniting all religions begins to break apart the moment one presses for details. For instance, the Abrahamic religions are incorrigibly dualistic and faith-based: In Judaism, Christianity, and Islam, the human soul is conceived as genuinely separate from the divine reality of God. The appropriate attitude for a creature that finds itself in this circumstance is some combination of terror, shame, and awe. In the best case, notions of God's love and grace provide some relief—but the central message of these faiths is that each of us is separate from, and in relationship to, a divine authority who will punish anyone who harbors the slightest doubt about His supremacy.

The Eastern tradition presents a very different picture of reality. And its highest teachings—found within the various schools of Buddhism and the nominally Hindu tradition of Advaita Vedanta—explicitly transcend dualism. By their lights, consciousness itself is identical to the very reality that one might otherwise mistake for God. While these teachings make metaphysical claims that any serious student of science should find incredible, they center on a range of experiences that the doctrines of Judaism, Christianity, and Islam rule out-of-bounds.

Of course, it is true that specific Jewish, Christian, and Muslim mystics have had experiences similar to those that motivate Buddhism and Advaita, but these contemplative insights are not exemplary of their faith. Rather, they are anomalies that Western mystics have always struggled to understand and to honor, often at considerable personal risk. Given their proper weight, these experiences produce heterodoxies for which Jews, Christians, and Muslims have been regularly exiled or killed.

Like Huxley, anyone determined to find a happy synthesis among spiritual traditions will notice that the Christian mystic Meister Eckhart (ca. 1260–ca. 1327) often sounded very much like a Buddhist: "The knower and the known are one. Simple people imagine that they should see God, as if He stood there and they here. This is not so. God and I, we are one in knowledge." But he also sounded like a man bound to be excommunicated by his church—as he was. Had Eckhart lived a little longer, it seems certain that he would have been dragged into the street and burned alive for these expansive ideas. That is a telling difference between Christianity and Buddhism.

In the same vein, it is misleading to hold up the Sufi mystic Al-Hallaj (858–922) as a representative of Islam. He was a Muslim, yes, but he suffered the most grisly death imaginable at the hands of his coreligionists for presuming to be one with God. Both Eckhart and Al-Hallaj gave voice to an experience of self-transcendence that any human being can, in principle, enjoy. However, their views were not consistent with the central teachings of their faiths.

The Indian tradition is comparatively free of problems of this kind. Although the teachings of Buddhism and Advaita are embedded in more or less conventional religions, they contain empirical insights about the nature of consciousness that do not depend upon faith. One can practice most techniques of Buddhist meditation or the method of self-inquiry of Advaita and experience the advertised changes in one's consciousness without ever believing in the law of karma or in the miracles attributed to Indian mystics. To get started as a Christian, however, one must first accept a dozen implausible things about the life of Jesus and the origins of the Bible—and the same can be said, minus a few unimportant details, about Judaism and Islam. If one should happen to discover that the sense of being an individual soul is an illusion, one will be guilty of blasphemy everywhere west of the Indus.

There is no question that many religious disciplines can produce interesting experiences in suitable minds. It should be clear, however, that engaging a faith-based (and probably delusional) practice, whatever its effects, isn't the same as investigating the nature of one's mind absent any doctrinal assumptions. Statements of this kind may seem starkly antagonistic toward Abrahamic religions, but they are nonetheless true: One can speak about Buddhism shorn of its miracles and irrational assumptions. The same cannot be said of Christianity or Islam.[3]

Western engagement with Eastern spirituality dates back at least as far as Alexander's campaign in India, where the young conqueror and his pet philosophers encountered naked ascetics whom they called "gymnosophists." It is often said that the thinking of these yogis greatly influenced the philosopher Pyrrho, the father of Greek skepticism. This seems a credible claim, because Pyrrho's teachings had much in common with Buddhism. But his contemplative insights and methods never became part of any system of thought in the West.

Serious study of Eastern thought by outsiders did not begin until the late eighteenth century. The first translation of a Sanskrit text into a Western language appears to have been Sir Charles Wilkins's rendering of the Bhagavad Gita, a cornerstone text of Hinduism, in 1785. The Buddhist canon would not attract the attention of Western scholars for another hundred years.[4]

The conversation between East and West started in earnest, albeit inauspiciously, with the birth of the Theosophical Society, that golem of spiritual hunger and self-deception brought into this world almost single-handedly by the incomparable Madame Helena Petrovna Blavatsky in 1875. Everything about Blavatsky seemed to defy earthly logic: She was an enormously fat woman who was said to have wandered alone and undetected for seven years in the mountains of Tibet. She was also thought to have survived shipwrecks, gunshot wounds, and sword fights. Even less persuasively, she claimed to be in psychic contact with members of the "Great White Brotherhood" of ascended masters—a collection of immortals responsible for the evolution and maintenance of the entire cosmos. Their leader hailed from the planet Venus but lived in the mythical kingdom of Shambhala, which Blavatsky placed somewhere in the vicinity of the Gobi Desert. With the suspiciously bureaucratic name "the Lord of the World," he supervised the work of other adepts, including the Buddha, Maitreya, Maha Chohan, and one Koot Hoomi, who appears to have had nothing better to do on behalf of the cosmos than to impart its secrets to Blavatsky.[5]

It is always surprising when a person attracts legions of followers and builds a large organization on their largesse while peddling penny-arcade mythology of this kind. But perhaps this was less remarkable in a time when even the best-educated people were still struggling to come to terms with electricity, evolution, and the existence of other planets. We can easily forget how suddenly the world had shrunk and the cosmos expanded as the nineteenth century came to a close. The geographical barriers between distant cultures had been stripped away by trade and conquest (one could now order a gin and tonic almost everywhere on earth), and yet the reality of unseen forces and alien worlds was a daily focus of the most careful scientific research. Inevitably, cross-cultural and scientific discoveries were mingled in the popular imagination with religious dogma and traditional occultism. In fact, this had been happening at the highest level of human thought for more than a century: It is always instructive to recall that the father of modern physics, Isaac Newton, squandered a considerable portion of his genius on the study of theology, biblical prophecy, and alchemy.

The inability to distinguish the strange but true from the merely strange was common enough in Blavatsky's time—as it is in our own. Blavatsky's contemporary Joseph Smith, a libidinous con man and crackpot, was able to found a new religion on the claim that he had unearthed the final revelations of God in the hallowed precincts of Manchester, New York, written in "reformed Egyptian" on golden plates. He decoded this text with the aid of magical "seer stones," which, whether by magic or not, allowed Smith to produce an English version of God's Word that was an embarrassing pastiche of plagiarisms from the Bible and silly lies about Jesus's life in America. And yet the resulting edifice of nonsense and taboo survives to this day.

A more modern cult, Scientology, leverages human credulity to an even greater degree: Adherents believe that human beings are possessed by the souls of extraterrestrials who were condemned to planet Earth 75 million years ago by the galactic overlord Xenu. How was their exile accomplished? The old-fashioned way: These aliens were shuttled by the billions to our humble planet aboard a spacecraft that resembled a DC-8. They were then imprisoned in a volcano and blasted to bits with hydrogen bombs. Their souls survived, however, and disentangling them from our own can be the work of a lifetime. It is also expensive.[6]

Despite the imponderables in her philosophy, Blavatsky was among the first people to announce in Western circles that there was such a thing as the "wisdom of the East." This wisdom began to trickle westward once Swami Vivekananda introduced the teachings of Vedanta at the World Parliament of Religions in Chicago in 1893. Again, Buddhism lagged behind: A few Western monks living on the island of Sri Lanka were beginning to translate the Pali Canon, which remains the most authoritative record of the teachings of the historical Buddha, Siddhartha Gautama. However, the practice of Buddhist meditation wouldn't actually be taught in the West for another half century.

It is easy enough to find fault with romantic ideas about Eastern wisdom, and a tradition of such criticism sprang up almost the instant the first Western seeker sat cross-legged and attempted to meditate. In the late 1950s, the author and journalist Arthur Koestler traveled to India and Japan in search of wisdom and summarized his pilgrimage thus: "I started my journey in sackcloth and ashes, and came back rather proud of being a European."[7]

In *The Lotus and the Robot*, Koestler gives some of his reasons for being less than awed by his journey to the East. Consider, for example, the ancient discipline of hatha yoga. While now generally viewed as a system of physical exercises designed to increase a person's strength and flexibility, in its traditional context hatha yoga is part of a larger effort to manipulate "subtle" features of the body unknown to anatomists. No doubt much of this subtlety corresponds to experiences that yogis actually have—but many of the beliefs formed on the basis of these experiences are patently absurd, and certain of the associated practices are both silly and injurious.

Koestler reports that the aspiring yogi is traditionally encouraged to lengthen his tongue—even going so far as to cut the frenulum (the membrane that anchors the tongue to the floor of the mouth) and stretch the soft palate. What is the purpose of these modifications? They enable our hero to insert his tongue into his nasopharynx, thereby blocking the flow of air through the nostrils. His anatomy thus improved, a yogi can then imbibe subtle liquors believed to emanate directly from his brain. These substances— imagined, by recourse to further subtleties, to be connected to the retention of semen—are said to confer not only spiritual wisdom but immortality. This technique of drinking mucus is known as *khechari mudra*, and it is thought to be one of the crowning achievements of yoga.

I'm more than happy to score a point for Koestler here. Needless to say, no defense of such practices will be found in this book.

Criticism of Eastern wisdom can seem especially pertinent when coming from Easterners themselves. There is indeed something preposterous about well-educated Westerners racing East in search of spiritual enlightenment while Easterners make the opposite pilgrimage seeking education and economic opportunities. I have a friend whose own adventures may have marked a high point in this global comedy. He made his first trip to India immediately after graduating from college, having already acquired several yogic affectations: He had the requisite beads and long hair, but he was also in the habit of writing the name of the Hindu god Ram in Devanagari script over and over in a journal. On the flight to the motherland, he had the good fortune to be seated next to an Indian businessman. This weary traveler thought he had witnessed every species of human folly—until he caught sight of my friend's scribbling. The spectacle of a Western-born Stanford graduate, of working age, holding degrees in both economics and history, devoting himself to the graphomaniacal worship of an imaginary deity in a language he could neither read nor understand was more than this man could abide in a confined space at 30,000 feet. After a testy exchange, the two travelers could only stare at each other in mutual incomprehension and pity—and they had ten hours yet to fly. There really are two sides to such a conversation, but I concede that only one of them can be made to look ridiculous.

We can also grant that Eastern wisdom has not produced societies or political institutions that are any better than their Western counterparts; in fact, one could argue that India has survived as the world's largest democracy only because of institutions that were built under British rule. Nor has the East led the world in scientific discovery. Nevertheless, there is something to the notion of uniquely Eastern wisdom, and most of it has been concentrated in or derived from the tradition of Buddhism.

Buddhism has been of special interest to Western scientists for reasons already hinted at. It isn't primarily a faith-based religion, and its central teachings are entirely empirical. Despite the superstitions that many Buddhists cherish, the doctrine has a practical and logical core that does not require any unwarranted assumptions. Many Westerners have recognized this and have been relieved to find a spiritual alternative to faith-based worship. It is no accident that most of the scientific research now done on meditation focuses primarily on Buddhist techniques.

Another reason for Buddhism's prominence among scientists has been the intellectual engagement of one of its most visible representatives: Tenzin Gyatso, the fourteenth Dalai Lama. Of course, the Dalai Lama is not without his critics. My late friend Christopher Hitchens meted out justice to "his holiness" on several occasions. He also castigated Western students of Buddhism for the "widely and lazily held belief that 'Oriental' religion is different from other faiths: less dogmatic, more contemplative, more . . . Transcendental," and for the "blissful, thoughtless exceptionalism" with which Buddhism is regarded by many.[8]

Hitch did have a point. In his capacity as the head of one of the four branches of Tibetan Buddhism and as the former leader of the Tibetan government in exile, the Dalai Lama has made some questionable claims and formed some embarrassing alliances. Although his engagement with science is far-reaching and surely sincere, the man is not above consulting an astrologer or "oracle" when making important decisions. I will have something to say in this book about many of the things that might have justified Hitch's opprobrium, but the general thrust of his commentary here was all wrong. Several Eastern traditions are exceptionally empirical and exceptionally wise, and therefore merit the exceptionalism claimed by their adherents.

Buddhism in particular possesses a literature on the nature of the mind that has no peer in Western religion or Western science. Some of these teachings are cluttered with metaphysical assumptions that should provoke our doubts, but many aren't. And when engaged as a set of hypotheses by which to investigate the mind and deepen one's ethical life, Buddhism can be an entirely rational enterprise.

Unlike the doctrines of Judaism, Christianity, and Islam, the teachings of Buddhism are not considered by their adherents to be the product of infallible revelation. They are, rather, empirical instructions: If you do X, you will experience Y. Although many Buddhists have a superstitious and cultic attachment to the historical Buddha, the teachings of Buddhism present him as an ordinary human being who succeeded in understanding the nature of his own mind. *Buddha* means "awakened one"—and Siddhartha Gautama was merely a man who woke up from the dream of being a separate self. Compare this with the Christian view of Jesus, who is imagined to be the son of the creator of the universe. This is a very different proposition, and it renders Christianity, no matter how fully divested of metaphysical baggage, all but irrelevant to a scientific discussion about the human condition.

The teachings of Buddhism, and of Eastern spirituality generally, focus on the primacy of the mind. There are dangers in this way of viewing the world, to be sure. Focusing on training the mind to the exclusion of all else can lead to political quietism and hive-like conformity. The fact that your mind is all you have and that it is possible to be at peace even in difficult circumstances can become an argument for ignoring obvious societal problems. But it is not a compelling one. The world is in desperate need of improvement—in global terms, freedom and prosperity remain the exception—and yet this doesn't mean we need to be miserable while we work for the common good.

In fact, the teachings of Buddhism emphasize a connection between ethical and spiritual life. Making progress in one domain lays a foundation for progress in the other. One can, for instance, spend long periods of time in contemplative solitude for the purpose of becoming a better person in the world—having better relationships, being more honest and compassionate and, therefore, more helpful to one's fellow human beings. Being wisely selfish and being selfless can amount to very much the same thing. There are centuries of anecdotal testimony on this point—and, as we will see, the scientific study of the mind has begun to bear it out. There is now little question that how one uses one's attention, moment to moment, largely determines what kind of person one becomes. Our minds—and lives—are largely shaped by how we use them.

Although the experience of self-transcendence is, in principle, available to everyone, this possibility is only weakly attested to in the religious and philosophical literature of the West. Only Buddhists and students of Advaita Vedanta (which appears to have been heavily influenced by Buddhism) have been absolutely clear in asserting that spiritual life consists in overcoming the illusion of the self by paying close attention to our experience in the present moment.[9]

As I wrote in my first book, *The End of Faith*, the disparity between Eastern and Western spirituality resembles that found between Eastern and Western medicine—with the arrow of embarrassment pointing in the opposite direction. Humanity did not understand the biology of cancer, develop antibiotics and vaccines, or sequence the human genome under an Eastern sun. Consequently, real medicine is almost entirely a product of Western science. Insofar as specific techniques of Eastern medicine actually work, they must conform, whether by design or by happenstance, to the principles of biology as we have come to know them in the West. This is not to say that Western medicine is complete. In a few decades, many of our current practices will seem barbaric. One need only ponder the list of side effects that accompany most medications to appreciate that these are terribly blunt instruments. Neverthe-

less, most of our knowledge about the human body—and about
the physical universe generally—emerged in the West. The rest is
instinct, folklore, bewilderment, and untimely death.

An honest comparison of spiritual traditions, Eastern and
Western, proves equally invidious. As manuals for contemplative
understanding, the Bible and the Koran are worse than useless.
Whatever wisdom can be found in their pages is never *best* found
there, and it is subverted, time and again, by ancient savagery and
superstition.

Again, one must deploy the necessary caveats: I am not saying that most Buddhists or Hindus have been sophisticated contemplatives. Their traditions have spawned many of the same
pathologies we see elsewhere among the faithful: dogmatism, anti-
intellectualism, tribalism, otherworldliness. However, the empirical
difference between the central teachings of Buddhism and Advaita
and those of Western monotheism is difficult to overstate. One
can traverse the Eastern paths simply by becoming interested in
the nature of one's own mind—especially in the immediate causes
of psychological suffering—and by paying closer attention to one's
experience in every present moment. There is, in truth, nothing
one need believe. The teachings of Buddhism and Advaita are best
viewed as lab manuals and explorers' logs detailing the results of
empirical research on the nature of human consciousness.

Nearly every geographical or linguistic barrier to the free exchange of ideas has now fallen away. It seems to me, therefore, that
educated people no longer have a right to any form of spiritual
provincialism. The truths of Eastern spirituality are now no more
Eastern than the truths of Western science are Western. We are
merely talking about human consciousness and its possible states.
My purpose in writing this book is to encourage you to investigate certain contemplative insights for yourself, without accepting

the metaphysical ideas that they inspired in ignorant and isolated peoples of the past.

A final word of caution: Nothing I say here is intended as a denial of the fact that psychological well-being requires a healthy "sense of self"—with all the capacities that this vague phrase implies. Children need to become autonomous, confident, and self-aware in order to form healthy relationships. And they must acquire a host of other cognitive, emotional, and interpersonal skills in the process of becoming sane and productive adults. Which is to say that there is a time and a place for everything—unless, of course, there isn't. No doubt there are psychological conditions, such as schizophrenia, for which practices of the sort I recommend in this book might be inappropriate. Some people find the experience of an extended, silent retreat psychologically destabilizing.[10] Again, an analogy to physical training seems apropos: Not everyone is suited to running a six-minute mile or bench-pressing his own body weight. But many quite ordinary people are capable of these feats, and there are better and worse ways to accomplish them. What is more, the same principles of fitness generally apply even to people whose abilities are limited by illness or injury.

So I want to make it clear that the instructions in this book are intended for readers who are adults (more or less) and free from any psychological or medical conditions that could be exacerbated by meditation or other techniques of sustained introspection. If paying attention to your breath, to bodily sensations, to the flow of thoughts, or to the nature of consciousness itself seems likely to cause you clinically significant anguish, please check with a psychologist or a psychiatrist before engaging in the practices I describe.

MINDFULNESS

It is always now. This might sound trite, but it is the truth. It's not quite true as a matter of neurology, because our minds are built upon layers of inputs whose timing we know must be different.[11] But it is true as a matter of *conscious experience.* The reality of your life is always now. And to realize this, we will see, is liberating. In fact, I think there is nothing more important to understand if you want to be happy in this world.

But we spend most of our lives forgetting this truth— overlooking it, fleeing it, repudiating it. And the horror is that we succeed. We manage to avoid being happy while struggling to *become* happy, fulfilling one desire after the next, banishing our fears, grasping at pleasure, recoiling from pain—and thinking, in- terminably, about how best to keep the whole works up and run- ning. As a consequence, we spend our lives being far less content than we might otherwise be. We often fail to appreciate what we have until we have lost it. We crave experiences, objects, relation- ships, only to grow bored with them. And yet the craving persists. I speak from experience, of course.

As a remedy for this predicament, many spiritual teachings ask us to entertain unfounded ideas about the nature of reality—or at the very least to develop a fondness for the iconography and rituals of one or another religion. But not all paths traverse the same rough ground. There are methods of meditation that do not require any artifice or unwarranted assumptions at all.

For beginners, I usually recommend a technique called *vipas- sana* (Pali for "insight"), which comes from the oldest tradition of Buddhism, the Theravada. One of the advantages of *vipassana* is that it can be taught in an entirely secular way. Experts in this practice generally acquire their training in a Buddhist context, and most retreat centers in the United States and Europe teach its associated Buddhist philosophy. Nevertheless, this method of introspection can be brought into any secular or scientific context without embarrassment. (The same cannot be said for the practice of chanting to Lord Krishna while banging a drum.) That is why *vipassana* is now being widely studied and adopted by psychologists and neuroscientists.

The quality of mind cultivated in *vipassana* is almost always referred to as "mindfulness," and the literature on its psychological benefits is now substantial. There is nothing spooky about mindfulness. It is simply a state of clear, nonjudgmental, and undistracted attention to the contents of consciousness, whether pleasant or unpleasant. Cultivating this quality of mind has been shown to reduce pain, anxiety, and depression; improve cognitive function; and even produce changes in gray matter density in regions of the brain related to learning and memory, emotional regulation, and self-awareness.[12] We will look more closely at the neurophysiology of mindfulness in a later chapter.

Mindfulness is a translation of the Pali word *sati*. The term has several meanings in the Buddhist literature, but for our purposes the most important is "clear awareness." The practice was first described in the *Satipatthana Sutta*,[13] which is part of the Pali Canon. Like many Buddhist texts, the *Satipatthana Sutta* is highly repetitive and, for anything but an avid student of Buddhism, exceptionally boring to read. However, when one compares texts of this kind with the Bible or the Koran, the difference is unmistakable: The *Satipatthana Sutta* is not a collection of ancient myths, superstitions, and taboos; it is a rigorously empirical guide to freeing the mind from suffering.

The Buddha described four foundations of mindfulness, which he taught as "the direct path for the purification of beings, for the surmounting of sorrow and lamentation, for the disappearance of pain and grief, for the attainment of the true way, for the realization of Nibbana" (Sanskrit, *Nirvana*). The four foundations of mindfulness are the body (breathing, changes in posture, activities), feelings (the senses of pleasantness, unpleasantness, and neutrality), the mind (in particular, its moods and attitudes), and the objects of mind (which include the five senses but also other mental states, such as volition, tranquility, rapture, equanimity, and even mindfulness itself). It is a peculiar list, at once redundant and incomplete—a problem that is compounded by the necessity of translating Pali terminology into English. The obvious message of the text, however, is that the totality of one's experience can become the field of contemplation. The meditator is merely instructed to pay attention, "ardently" and "fully aware" and "free from covetousness and grief for the world."

There is nothing passive about mindfulness. One might even say that it expresses a specific kind of passion—a passion for discerning what is subjectively real in every moment. It is a mode of cognition that is, above all, undistracted, accepting, and (ultimately) nonconceptual. Being mindful is not a matter of *thinking* more clearly about experience; it is the act of *experiencing* more clearly, including the arising of thoughts themselves. Mindfulness is a vivid awareness of whatever is appearing in one's mind or body—thoughts, sensations, moods—without grasping at the pleasant or recoiling from the unpleasant. One of the great strengths of this technique of meditation, from a secular point of view, is that it does not require us to adopt any cultural affectations or unjustified beliefs. It simply demands that we pay close attention to the flow of experience in each moment.

The principal enemy of mindfulness—or of any meditative

practice—is our deeply conditioned habit of being distracted by thoughts. The problem is not thoughts themselves but the state of thinking without knowing that we are thinking. In fact, thoughts of all kinds can be perfectly good objects of mindfulness. In the early stages of one's practice, however, the arising of thought will be more or less synonymous with distraction—that is, with a failure to meditate. Most people who believe they are meditating are merely thinking with their eyes closed. By practicing mindfulness, however, one can awaken from the dream of discursive thought and begin to see each arising image, idea, or bit of language vanish without a trace. What remains is consciousness itself, with its attendant sights, sounds, sensations, and thoughts appearing and changing in every moment.

In the beginning of one's meditation practice, the difference between ordinary experience and what one comes to consider "mindfulness" is not very clear, and it takes some training to distinguish between being lost in thought and seeing thoughts for what they are. In this sense, learning to meditate is just like acquiring any other skill. It takes many thousands of repetitions to throw a good jab or to coax music from the strings of a guitar. With practice, mindfulness becomes a well-formed habit of attention, and the difference between it and ordinary thinking will become increasingly clear. Eventually, it begins to seem as if you are repeatedly awakening from a dream to find yourself safely in bed. No matter how terrible the dream, the relief is instantaneous. And yet it is difficult to stay awake for more than a few seconds at a time.

My friend Joseph Goldstein, one of the finest *vipassana* teachers I know, likens this shift in awareness to the experience of being fully immersed in a film and then suddenly realizing that you are sitting in a theater watching a mere play of light on a wall. Your perception is unchanged, but the spell is broken. Most of us spend every waking moment lost in the movie of our lives. Until we see that an alternative to this enchantment exists, we are entirely at the mercy of appearances. Again, the difference I am describing is not a matter of achieving a new conceptual understanding or of adopting new beliefs about the nature of reality. The change comes when we experience the present moment prior to the arising of thought.

The Buddha taught mindfulness as the appropriate response to the truth of *dukkha*, usually translated from the Pali, somewhat misleadingly, as "suffering." A better translation would be "unsatisfactoriness." Suffering may not be inherent in life, but unsatisfactoriness is. We crave lasting happiness in the midst of change: Our bodies age, cherished objects break, pleasures fade, relationships fail. Our attachment to the good things in life and our aversion to the bad amount to a denial of these realities, and this inevitably leads to feelings of dissatisfaction. Mindfulness is a technique for achieving equanimity amid the flux, allowing us to simply be aware of the quality of experience in each moment, whether pleasant or unpleasant. This may seem like a recipe for apathy, but it needn't be. It is actually possible to be mindful—and, therefore, to be at peace with the present moment—even while working to change the world for the better.

Mindfulness meditation is extraordinarily simple to describe, but it isn't easy to perform. True mastery might require special talent and a lifetime of devotion to the task, and yet a genuine transformation in one's perception of the world is within reach for most of us. Practice is the only thing that will lead to success. The simple instructions given in the box that follows are analogous to instructions on how to walk a tightrope—which, I assume, must go something like this:

1. Find a horizontal cable that can support your weight.
2. Stand on one end.
3. Step forward by placing one foot directly in front of the other.
4. Repeat.
5. Don't fall.

Clearly, steps 2 through 5 entail a little trial and error. Happily, the benefits of training in meditation arrive long before mastery does. And falling, for our purposes, occurs almost ceaselessly, every time we become lost in thought. Again, the problem is not thoughts themselves but the state of thinking *without being fully aware that we are thinking*.

As every meditator soon discovers, distraction is the normal condition of our minds: Most of us topple from the wire every second—whether gliding happily into reverie or plunging into fear, anger, self-hatred, and other negative states of mind. Meditation is a technique for waking up. The goal is to come out of the trance of discursive thinking and to stop reflexively grasping at the pleasant and recoiling from the unpleasant, so that we can enjoy a mind undisturbed by worry, merely open like the sky, and effortlessly aware of the flow of experience in the present.

How to Meditate

1. Sit comfortably, with your spine erect, either in a chair or cross-legged on a cushion.
2. Close your eyes, take a few deep breaths, and feel the points of contact between your body and the chair or the floor. Notice the sensations associated with sitting— feelings of pressure, warmth, tingling, vibration, etc.

3. Gradually become aware of the process of breathing. Pay attention to wherever you feel the breath most distinctly—either at your nostrils or in the rising and falling of your abdomen.

4. Allow your attention to rest in the mere sensation of breathing. (You don't have to control your breath. Just let it come and go naturally.)

5. Every time your mind wanders in thought, gently return it to the breath.

6. As you focus on the process of breathing, you will also perceive sounds, bodily sensations, or emotions. Simply observe these phenomena as they appear in consciousness and then return to the breath.

7. The moment you notice that you have been lost in thought, observe the present thought itself as an object of consciousness. Then return your attention to the breath—or to any sounds or sensations arising in the next moment.

8. Continue in this way until you can merely witness all objects of consciousness—sights, sounds, sensations, emotions, even thoughts themselves—as they arise, change, and pass away.

Those who are new to this practice generally find it useful to hear instructions of this kind spoken aloud during the course of a meditation session. I have posted guided meditations of varying length on my website.

THE TRUTH OF SUFFERING

I am sitting in a coffee shop in midtown Manhattan, drinking exactly what I want (coffee), eating exactly what I want (a cookie), and doing exactly what I want (writing this book). It is a beautiful fall day, and many of the people passing by on the sidewalk appear to radiate good fortune from their pores. Several are so physically attractive that I'm beginning to wonder whether Photoshop can now be applied to the human body. Up and down this street, and for a mile in each direction, stores sell jewelry, art, and clothing that not even 1 percent of humanity could hope to purchase.

So what did the Buddha mean when he spoke of the "unsatisfactoriness" (*dukkha*) of life? Was he referring merely to the poor and the hungry? Or are these rich and beautiful people suffering even now? Of course, suffering is all around us—even here, where everything appears to be going well for the moment.

First, the obvious: Within a few blocks of where I am sitting are hospitals, convalescent homes, psychiatrists' offices, and other rooms built to assuage, or merely to contain, some of the most profound forms of human misery. A man runs over his own child while backing his car out of the driveway. A woman learns that she has terminal cancer on the eve of her wedding. We know that the worst can happen to anyone at any time—and most people spend a great deal of mental energy hoping that it won't happen to them.

But more subtle forms of suffering can be found, even among people who seem to have every reason to be satisfied in the present. Although wealth and fame can secure many forms of pleasure, few of us have any illusions that they guarantee happiness. Anyone who owns a television or reads the newspaper has seen movie stars, politicians, professional athletes, and other celebrities ricochet from marriage to marriage and from scandal to scandal. To learn that a young, attractive, talented, and successful person is nevertheless addicted to drugs or clinically depressed is to be given almost no cause for surprise.

Yet the unsatisfactoriness of the good life runs deeper than this.

Even while living safely between emergencies, most of us feel a wide range of painful emotions on a daily basis. When you wake up in the morning, are you filled with joy? How do you feel at work or when looking in the mirror? How satisfied are you with what you've accomplished in life? How much of your time with your family is spent surrendered to love and gratitude, and how much is spent just struggling to be happy in one another's company? Even for extraordinarily lucky people, life is difficult. And when we look at what makes it so, we see that we are all prisoners of our thoughts.

And then there is death, which defeats everyone. Most people seem to believe that we have only two ways to think about death: We can fear it and do our best to ignore it, or we can deny that it is real. The first strategy leads to a life of conventional worldliness and distraction—we merely strive for pleasure and success and do our best to keep the reality of death out of view. The second strategy is the province of religion, which assures us that death is but a doorway to another world and that the most important opportunities in life occur after the lifetime of the body. But there is another path, and it seems the only one compatible with intellectual honesty. That path is the subject of this book.

ENLIGHTENMENT

What is enlightenment, which is so often said to be the ultimate goal of meditation? There are many esoteric details that we can safely ignore—disagreements among contemplative traditions about what, exactly, is gained or lost at the end of the spiritual path. Many of these claims are preposterous. Within most schools of Buddhism, for instance, a buddha—whether the historical Buddha, Siddhartha Gautama, or any other person who attains the state of "full enlightenment"—is generally described as "omniscient." Just what this means is open to a fair bit of caviling. But however narrowly defined, the claim is absurd. If the historical Buddha were "omniscient," he would have been, at minimum, a better mathematician, physicist, biologist, and *Jeopardy* contestant than any person who has ever lived. Is it reasonable to expect that an ascetic in the fifth century BC, by virtue of his meditative insights, spontaneously became an unprecedented genius in every field of human inquiry, including those that did not exist at the time in which he lived? Would Siddhartha Gautama have awed Kurt Gödel, Alan Turing, John von Neumann, and Claude Shannon with his command of mathematical logic and information theory? Of course not. To think otherwise is pure, religious piety.

Any extension of the notion of "omniscience" to procedural knowledge—that is, to knowing how to *do* something—would render the Buddha capable of painting the Sistine Chapel in the morning and demolishing Roger Federer at Centre Court in the afternoon. Is there any reason to believe that Siddhartha Gautama, or any other celebrated contemplative, possessed such abilities by virtue of his spiritual practice? None whatsoever. Nevertheless, many Buddhists believe that buddhas can do all these things and more. Again, this is religious dogmatism, not a rational approach to spiritual life.[14]

I make no claims in support of magic or miracles in this book. However, I can say that the true goal of meditation is more profound than most people realize—and it does, in fact, encompass many of the experiences that traditional mystics claim for themselves. It is quite possible to lose one's sense of being a separate self and to experience a kind of boundless, open awareness—to feel, in other words, at one with the cosmos. This says a lot about the possibilities of human consciousness, but it says nothing about the universe at large. And it sheds no light at all on the relationship between mind and matter. The fact that it is possible to love one's neighbor as oneself should be a great finding for the field of psychology, but it lends absolutely no credence to the claim that Jesus was the son of God, or even that God exists. Nor does it suggest that the "energy" of love somehow pervades the cosmos. These are historical and metaphysical claims that personal experience cannot justify.

However, a phenomenon like self-transcending love does entitle us to make claims about the human mind. And this particular experience is so well attested and so readily achieved by those who devote themselves to specific practices (the Buddhist technique of *metta* meditation, for instance) or who even take the right drug (MDMA) that there is very little controversy that it exists. Facts of this kind must now be understood in a rational context.

The traditional goal of meditation is to arrive at a state of well-being that is imperturbable—or if perturbed, easily regained. The French monk Matthieu Ricard describes such happiness as "a deep sense of flourishing that arises from an exceptionally healthy mind."[15] The purpose of meditation is to recognize that you already have such a mind. That discovery, in turn, helps you to cease doing the things that produce needless confusion and suffering for yourself and others. Of course, most people never truly master the practice and don't reach a condition of imperturbable happiness. The near goal, therefore, is to have an *increasingly* healthy mind— that is, to be moving one's mind in the right direction.

There is nothing novel about trying to *become* happy. And one *can* become happy, within certain limits, without any recourse to the practice of meditation. But conventional sources of happiness are unreliable, being dependent upon changing conditions. It is difficult to raise a happy family, to keep yourself and those you love healthy, to acquire wealth and find creative and fulfilling ways to enjoy it, to form deep friendships, to contribute to society in ways that are emotionally rewarding, to perfect a wide variety of artistic, athletic, and intellectual skills—and to keep the machinery of happiness running day after day. There is nothing wrong with being fulfilled in all these ways—except for the fact that, if you pay close attention, you will see that there is still something wrong with it. These forms of happiness aren't good enough. Our feelings of fulfillment do not last. And the stress of life continues.

So what would a spiritual master be a master *of*? At a minimum, she will no longer suffer certain cognitive and emotional illusions—above all, she will no longer feel identical to her thoughts. Once again, this is not to say that such a person will no longer think, but she would no longer succumb to the primary confusion that thoughts produce in most of us: She would no longer feel that there is an inner self who is a thinker of these thoughts. Such a person will naturally maintain an openness and serenity of mind that is available to most of us only for brief moments, even after years of practice. I remain agnostic as to whether anyone has achieved such a state permanently, but I know from direct experience that it is possible to be far more enlightened than I tend to be.

The question of whether enlightenment is a permanent state need not detain us. The crucial point is that you can glimpse something about the nature of consciousness that will liberate you from suffering in the present. Even just recognizing the impermanence of your mental states—deeply, not merely as an idea—can transform your life. Every mental state you have ever had has arisen and then passed away. This is a first-person fact—but it is, nonetheless, a fact that any human being can readily confirm. We don't have to know any more about the brain or about the relationship between consciousness and the physical world to understand this truth about our own minds. The promise of spiritual life—indeed, the very thing that makes it "spiritual" in the sense I invoke throughout this book—is that there are truths about the mind that we are better off knowing. What we need to become happier and to make the world a better place is not more pious illusions but a clearer understanding of the way things are.

The moment we admit the possibility of attaining contemplative insights—and of training one's mind for that purpose—we must acknowledge that people naturally fall at different points on a continuum between ignorance and wisdom. Part of this range will be considered "normal," but normal isn't necessarily a happy place to be. Just as a person's physical body and abilities can be refined—Olympic athletes are *not* normal—one's mental life can deepen and expand on the basis of talent and training. This is nearly self-evident, but it remains a controversial point. No one hesitates to admit the role of talent and training in the context of physical and intellectual pursuits; I have never met another person who denied that some of us are stronger, more athletic, or more learned than others. But many people find it difficult to acknowledge that a continuum of moral and spiritual wisdom exists or that there might be better and worse ways to traverse it.

Stages of spiritual development, therefore, appear unavoidable. Just as we must grow into adulthood physically—and we can fail to mature or become sick or injured along the way—our minds develop by degrees. One can't learn sophisticated skills such as syllogistic reasoning, algebra, or irony until one has acquired more basic skills. It seems to me that a healthy spiritual life can begin only once our physical, mental, social, and ethical lives have sufficiently matured. We must learn to use language before we can work with it creatively or understand its limits, and the conventional self must form before we can investigate it and understand that it is not what it appears to be. An ability to examine the contents of one's own consciousness clearly, dispassionately, and nondiscursively, with sufficient attention to realize that no inner self exists, is a very sophisticated skill. And yet basic mindfulness can be practiced very early in life. Many people, including my wife, have successfully taught it to children as young as six. At that age—and every age thereafter—it can be a powerful tool for self-regulation and self-awareness.

Contemplatives have long understood that positive habits of mind are best viewed as skills that most of us learn imperfectly as we grow to adulthood. It is possible to become more focused, patient, and compassionate than one naturally tends to be, and there are many things to learn about how to be happy in this world. These are truths that Western psychological science has only recently begun to explore.

Some people are content in the midst of deprivation and danger, while others are miserable despite having all the luck in the world. This is not to say that external circumstances do not matter. But it is your mind, rather than circumstances themselves, that determines the quality of your life. Your mind is the basis of everything you experience and of every contribution you make to the lives of others. Given this fact, it makes sense to train it.

Scientists and skeptics generally assume that the traditional claims of yogis and mystics must be exaggerated or simply delusional and that the only rational purpose of meditation is limited to conventional "stress reduction." Conversely, serious students of these practices often insist that even the most outlandish claims made by and about spiritual masters are true. I am attempting to lead the reader along a middle path between these extremes—one that preserves our scientific skepticism but acknowledges that it is possible to radically transform our minds.

In one sense, the Buddhist concept of enlightenment really is just the epitome of "stress reduction"—and depending on how much stress one reduces, the results of one's practice can seem more or less profound. According to the Buddhist teachings, human beings have a distorted view of reality that leads them to suffer unnecessarily. We grasp at transitory pleasures. We brood about the past and worry about the future. We continually seek to prop up and defend an egoic self that doesn't exist. This is stressful—and spiritual life is a process of gradually unraveling our confusion and bringing this stress to an end. According to the Buddhist view, by seeing things as they are, we cease to suffer in the usual ways, and our minds can open to states of well-being that are intrinsic to the nature of consciousness.

Of course, some people claim to love stress and appear eager to live by its logic. Some even derive pleasure from imposing stress on others. Genghis Khan is reported to have said, "The greatest happiness is to scatter your enemy and drive him before you, to see his cities reduced to ashes, to see those who love him shrouded in tears, and to gather to your bosom his wives and daughters." People attach many meanings to terms like *happiness*, and not all of them are compatible with one another.

In *The Moral Landscape*, I argued that we tend to be unnecessarily confused by differences of opinion on the topic of human well-being. No doubt certain people can derive mental pleasure—and even experience genuine ecstasy—by behaving in ways that produce immense suffering for others. But we know that these states are anomalous—or, at least, not sustainable—because we depend upon one another for more or less everything. Whatever the associated pleasures, raping and pillaging can't be a stable strategy for finding happiness in this world. Given our social requirements, we know that the deepest and most durable forms of well-being must be compatible with an ethical concern for other people—even for complete strangers—otherwise, violent conflict becomes inevitable. We also know that there are certain forms of happiness that are not available to a person even if, like Genghis Khan, he finds himself on the winning side of every siege. Some pleasures are intrinsically ethical—feelings like love, gratitude, devotion, and compassion. To inhabit these states of mind is, by definition, to be brought into alignment with others.

In my view, the realistic goal to be attained through spiritual practice is not some permanent state of enlightenment that admits of no further efforts but a capacity to be free in this moment, in the midst of whatever is happening. If you can do that, you have already solved most of the problems you will encounter in life.

Chapter 2

The Mystery of Consciousness

I nvestigating the nature of consciousness itself—and trans-forming its contents through deliberate training—is the basis of spiritual life. In scientific terms, however, consciousness remains notoriously difficult to understand, or even to define. In fact, many debates about its character have been waged without the participants' finding even a common topic as common ground. While we need not recapitulate the history of our confusion on this point, it will be useful to briefly examine why consciousness still poses a unique challenge to science. Having done so, we will see that spirituality is not just important for living a good life; it is actually essential for understanding the human mind.

In one of the most influential essays on consciousness ever written, the philosopher Thomas Nagel asks us to consider what it is like to be a bat.[1] His interest isn't in bats but in how we define the concept of "consciousness." Nagel argues that an organism is conscious "if and only if there is something that it is like to *be* that organism—something that it is like *for* the organism." Whether you find that statement brilliant, trivial, or merely perplexing probably says a lot about your appetite for philosophy. "Brilliant" and "trivial" can both be defended, but Nagel's claim needn't leave you confused. He is simply asking you to imagine trading places with a bat. If you would be left with *any* experience, however indescribable—some spectrum of sights, sounds, sensations, feelings—*that* is what consciousness is in the case of a bat. If being transformed into a bat were tantamount to annihilation, however, then bats are not conscious.[2] Nagel's point is that whatever else consciousness may or may not entail in physical terms, the difference between it and unconsciousness is a matter of subjective experience. Either the lights are on, or they are not.[3]

But experience is one thing, and our growing scientific picture of reality is another. At this moment, you might be vividly aware of reading this book, but you are completely unaware of the electrochemical events occurring at each of the trillions of synapses in your brain. However much you may know about physics, chemistry, and biology, you live elsewhere. As a matter of your *experience*, you are not a body of atoms, molecules, and cells; you are consciousness and its ever-changing contents, passing through various stages of wakefulness and sleep, from cradle to grave.

And the question of how consciousness relates to the physical world remains famously unresolved. There are reasons to believe that it emerges on the basis of information processing in complex systems like a human brain, because when we look at the universe, we find it filled with simpler structures, like stars, and processes, like nuclear fusion, that offer no outward signs of consciousness. But our intuitions here may not amount to much. After all, how would the sun appear if it were conscious? Perhaps exactly as it does now. (Would you expect it to talk?) And yet somehow it seems far less likely that the stars are conscious and simply mute than that they lack inner lives altogether.

Whatever the ultimate relationship between consciousness and matter, almost everyone will agree that at some point in the development of complex organisms like ourselves, consciousness *seems* to emerge. This emergence does not depend on a change of materials, for you and I are built of the same atoms as a fern or a ham sandwich. Instead, the birth of consciousness must be the result of organization: Arranging atoms in certain ways appears to bring about an experience of *being* that very collection of atoms. This is undoubtedly one of the deepest mysteries given to us to contemplate.[4]

Nevertheless, Nagel was right to observe that the reality of consciousness is, first and foremost, subjective—for it is simply the fact of subjectivity itself. And whether something *seems* conscious from the outside is never quite the point. I happen to know a person who once woke up during a surgery for which he had received a general anesthetic. Owing to the paralytic component of the anesthesia, however, he was unable to signal to his doctors that he was awake and feeling rather more of the procedure than he liked. This was inconvenient, to say the least, because they were in the process of replacing his liver. If you think the important part of consciousness is its link to speech and behavior, spare a moment to consider the problem of "anesthesia awareness." It is a cure for much bad philosophy.[5]

It is surely a sign of intellectual progress that a discussion of consciousness need no longer begin with a debate about its existence. To say that consciousness may only *seem* to exist, from the inside, is to admit its existence in full—for if things seem any way at all, *that* is consciousness. Even if I happen to be a brain in a vat at this moment—and all my memories are false, and all my perceptions are of a world that does not exist—the fact that I am *having an experience* is indisputable (to me, at least). This is all that is required for me (or any other sentient being) to fully establish the reality of consciousness. Consciousness is the one thing in this universe that cannot be an illusion.[6]

As our understanding of the physical world has evolved, our notion of what counts as "physical" has broadened considerably. A world teeming with fields and forces, vacuum fluctuations, and the other gossamer spawn of modern physics is not the physical world of common sense. In fact, our common sense seems to be stuck somewhere in the sixteenth century. It has also been generally forgotten that many of the patriarchs of physics in the first half of the twentieth century regularly impugned the "physicality" of the universe and placed mind—or thoughts, or consciousness itself—at the very wellspring of reality. Nonreductive views like those of Arthur Eddington, James Jeans, Wolfgang Pauli, Werner Heisenberg, and Erwin Schrödinger seem to have had no lasting impact.[7] In some ways we can be thankful for this, for a fair amount of mumbo jumbo was in the air. Pauli, for instance, was a devotee of Carl Jung, who apparently analyzed no fewer than 1,300 of the great man's dreams.[8] Although Pauli was one of the titans of physics, his thoughts about the irreducibility of mind probably had as much to do with Jung's febrile imagination as they did with quantum mechanics.

The allure of the numinous eventually subsided. Once physicists got down to the serious business of building bombs, we were apparently returned to a universe of objects—and to a style of discourse, across all branches of science and philosophy, that made the mind seem ripe for reduction to the "physical" world.

These developments have greatly inconvenienced New Age thinkers—or would have, had they deigned to notice them.

Authors struggling to link spirituality to science generally pin their hopes on misunderstandings of the "Copenhagen interpretation of quantum mechanics," which they take as proof that consciousness plays a central role in determining the character of the physical world. If nothing is real until it is observed, consciousness cannot arise from electrochemical events in the brains of animals like ourselves; rather, it must be part of the very fabric of reality. But this simply isn't the position of mainstream physics. It is true that, according to Copenhagen, quantum mechanical systems do not behave classically until they are observed, and before that they may seem to exist in many different states simultaneously. But what counts as "observation" under the original Copenhagen view was never clearly defined. The notion has been refined since, and it has nothing to do with consciousness. It's not that the mysteries of quantum mechanics have been resolved—the physical picture is strange however one looks at it. And the problem of how an underlying quantum mechanical reality becomes the seemingly classical world of tables and chairs hasn't been completely understood. However, there is no reason to think that consciousness is integral to the process. It seems certain, therefore, that anyone who would base his spirituality on misinterpretations of 1930s physics is bound to be disappointed. As we will see, the link between spirituality and science must be found in another place.[9]

We know, of course, that human *minds* are the product of human brains. There is simply no question that your ability to decode and understand this sentence depends upon neurophysiological events taking place inside your head at this moment. But most of this mental work occurs entirely in the dark, and it is a mystery why any part of the process should be attended by consciousness.

Nothing about a brain, when surveyed as a physical system, suggests that it is a locus of experience. Were we not already brimming with consciousness ourselves, we would find no evidence for it in the universe—nor would we have any notion of the many experiential states that it gives rise to. The only proof that it is like something to be you at this moment is the fact (obvious only to you) that it is like something to be you.[10]

However we propose to explain the emergence of consciousness—be it in biological, functional, computational, or any other terms—we have committed ourselves to this much: First there is a physical world, unconscious and seething with unperceived events; then, by virtue of some physical property or process, consciousness itself springs, or staggers, into being. This idea seems to me not merely strange but perfectly mysterious. That doesn't mean it isn't true. When we linger over the details, however, this notion of emergence seems merely a placeholder for a miracle.

Consciousness—the sheer fact that this universe is illuminated by sentience—is precisely what unconsciousness is not. And I believe that no description of unconscious complexity will fully account for it. To simply assert that consciousness arose at some point in the evolution of life, and that it results from a specific arrangement of neurons firing in concert within an individual brain, doesn't give us any inkling of *how* it could emerge from unconscious processes, even in principle. However, this is not to say that some other thesis about consciousness must be true. Consciousness may very well be the lawful product of unconscious information processing. But I don't know what that sentence actually means—and I don't think anyone else does either.[11] This situation has been characterized as an "explanatory gap"[12] and as the "hard problem of consciousness,"[13] and it is surely both. Some philosophers have suggested that the relationship between mind and body will be understood only with reference to concepts that are neither physical nor mental but that are in some way "neutral."[14] Others claim that consciousness can be known to be the product of physical causes but cannot be conceptually reduced to such causes.[15] Still others have argued that the notion of a nonreductive physical account is incoherent.[16]

I am sympathetic with those who, like the philosopher Colin McGinn and the psychologist Steven Pinker, have suggested that perhaps the emergence of consciousness is simply incomprehensible in human terms.[17] Every chain of explanation must end somewhere—generally with a brute fact that neglects to explain itself. Perhaps consciousness presents an impasse of this sort.[18]

In any case, the task of explaining consciousness in physical terms bears little resemblance to other successful explanations in the history of science. The analogies that scientists and philosophers marshal here are invariably misleading. The fact, for instance, that we can now describe the properties of matter, such as fluidity, in terms of microscopic events that are not themselves "fluid" does not suggest a way to understand consciousness as an emergent property of the unconscious world. It is easy to see that no single water molecule can be "fluid," and it is easy to see that billions of such molecules, freely sliding past one another, would appear as "fluidity" on the scale of a human hand. What is not easy to see is how analogies of this kind have persuaded so many people that consciousness can be readily explained in terms of information processing.[19]

For an explanation of a phenomenon to be satisfying, it must first be, at a minimum, *intelligible*. In this regard, the emergence of fluidity poses no problems: The free sliding of molecules seems exactly the sort of thing that should be true of a substance to ensure its fluidity. Why can I pass my hand through liquid water and not through rock? Because the molecules of water are not bound so tightly as to resist my motion. Notice that this explanation of fluidity is perfectly reductive: Fluidity really is "nothing but" the free motion of molecules. For this explanation to be sufficient, we must admit that molecules exist, of course, but once we do, the problem is solved. No one has described a set of unconscious events whose sufficiency as a cause of consciousness would make sense in this way. Any attempt to understand consciousness in terms of brain activity merely correlates a person's ability to report an experience (demonstrating that he was aware of it) with specific states of his brain. While such correlations can amount to fascinating neuroscience, they bring us no closer to explaining the emergence of consciousness itself.

There will almost certainly come a time when we will build a robot whose facial expressiveness, tone of voice, and flexibility of thought will cause us to wonder whether or not it is conscious. This robot might even claim to be conscious and be eager to participate in the kinds of experiments we now perform on human beings, allowing us to correlate its responses to stimuli with changes in its "brain." It seems clear, however, that unless we can do more than this, we will never know whether there is "something that it is like" to be such a machine.[20]

Some readers may think that I've stacked the deck against the sciences of the mind by comparing consciousness to a phenomenon as easily understood as fluidity. Surely science has dispelled far greater mysteries. What, for instance, is the difference between a living system and a dead one? Insofar as questions about consciousness itself can be kept off the table, it seems that the difference is now reasonably clear to us. And yet, as late as 1932, the Scottish physiologist J. S. Haldane (father of J. B. S. Haldane) wrote:

What intelligible account can the mechanistic theory of life give of the . . . recovery from disease and injuries? Simply none at all, except that these phenomena are so complex and strange that as yet we cannot understand them. It is exactly the same with the closely related phenomena of reproduction. We cannot by any stretch of the imagination conceive a delicate and complex mechanism which is capable, like a living organism, of reproducing itself indefinitely often.[21]

Scarcely twenty years passed before our imaginations were duly stretched. Much work in biology remains to be done, but anyone who entertains *vitalism** at this point is simply ignorant about the nature of living systems. The jury is no longer out on questions of this kind, and more than half a century has passed since the earth's creatures required an *élan vital* to propagate themselves or to recover from injury. Is my skepticism that we will arrive at a physical explanation of consciousness analogous to Haldane's doubt about the feasibility of explaining life in terms of processes that are not themselves alive?

It wouldn't seem so. To say that a system is alive is very much like saying that it is fluid, because life is a matter of what systems do with respect to their environment. Like fluidity, life is defined according to external criteria. Consciousness is not (and, I think, cannot be). We would never have occasion to say of something that does not eat, excrete, grow, or reproduce that it might be "alive." It might, however, be conscious.[22]

Might a mature neuroscience nevertheless offer a proper explanation of consciousness in terms of its underlying brain processes?

---

* *Vitalism* is the now discredited doctrine that living systems require some nonphysical principle to explain their organization and behavior.

Again, there is nothing about a brain, studied at any scale, that even *suggests* that it might harbor consciousness—apart from the fact that we experience consciousness directly and have correlated many of its contents, or lack thereof, with processes in our brains. Nothing about human behavior or language or culture demonstrates that it is mediated by consciousness, apart from the fact that we simply know that it is—a truth that someone can appreciate in himself directly and in others by analogy.[23]

Here is where the distinction between studying consciousness itself and studying its *contents* becomes paramount. It is easy to see how the contents of consciousness might be understood in neurophysiological terms. Consider, for instance, our experience of seeing an object: Its color, contours, apparent motion, and location in space arise in consciousness as a seamless unity, even though this information is processed by many separate systems in the brain. Thus, when a golfer prepares to hit a shot, he does not first see the ball's roundness, then its whiteness, and only then its position on the tee. Rather, he enjoys a unified perception of the ball. Many neuroscientists believe that this phenomenon of "binding" can be explained by disparate groups of neurons firing in synchrony.[24] Whether or not this theory is true, it is at least intelligible—because synchronous activity seems just the sort of thing that could explain the unity of a percept.

This work suggests, as many other findings in neuroscience do, that the *contents* of consciousness can often be made sense of in terms of their underlying neurophysiology.[25] However, when we ask why such phenomena should be experienced in the first place, we are returned to the mystery of consciousness in full.[26]

Unfortunately, efforts to locate consciousness in the brain generally fail to distinguish between consciousness and its contents. As a result, many researchers have taken one form of consciousness (or one class of its contents) as a sufficient view of the whole. For instance, Christof Koch and others have done some very clever work on vision, looking for which regions of the brain encode conscious visual perception.[27] The phenomenon of *binocular rivalry* has provided an especially useful foothold here: It just so happens that when each eye is presented with a different visual stimulus, a person's conscious experience is not a blending of the two images but, rather, a series of apparently random transitions between them. If, for instance, you are shown a picture of a house in one eye and a human face in the other, you will not see the two images competing with each other or otherwise superimposed. You will see the house for a few seconds, and then the face, and then the house again, switching at random intervals. This phenomenon has allowed experimenters to look for those regions of the brain (in both humans and monkeys) that respond to a change in conscious perception. The psychophysical situation seems tailor-made to distinguish the frontier between the conscious and unconscious components of vision, because the input remains constant—each eye receives the continuous impression of a single image—while somewhere in the brain a wholesale change in the contents of consciousness occurs every few seconds. This is very interesting—and yet subjects experiencing binocular rivalry are *conscious* throughout the experiment; only the contents of visual awareness have been modulated by the task. If you shut your eyes at this moment, the contents of your consciousness change quite drastically, but your consciousness (arguably) does not.

This is not to say that our understanding of the mind won't change in surprising ways through our study of the brain. There may be no limit to how a maturing neuroscience might reshape our beliefs about the nature of conscious experience. Are we unconscious during sleep or merely unable to remember what sleep is like? Can human minds be duplicated? Neuroscience may one day answer such questions—and the answers might well surprise us.

But the reality of consciousness appears irreducible. Only consciousness can know itself—and directly, through first-person experience. It follows, therefore, that rigorous introspection—"spirituality" in the widest sense of the term—is an indispensable part of understanding the nature of the mind.

THE MIND DIVIDED

If spirituality is to become part of science, however, it must integrate with the rest of what we know about the world. It has long been obvious that traditional approaches to spirituality cannot do this—being based, to one or another degree, on religious myths and superstitions. Consider the idea that human beings, alone among Nature's animals, have been installed with immortal souls. This dogma came under pressure the moment Darwin published *On the Origin of Species* in 1859, but it is now truly dead. By sequencing a wide variety of genomes, we have finally rendered our continuity with the rest of life undeniable. We are such stuff as yeasts are made of. Of course, only 25 percent of Americans believe in evolution (while 68 percent believe in the literal existence of Satan).[28] But we can now say that any conception of our place in the universe that denies we evolved from more primitive life forms is pure delusion.

Neuroscience has also produced results that are equally hostile to the traditional idea of souls—and, therefore, to any approach to spirituality that presupposes their existence. One such finding, conclusively demonstrated in humans and animals since the 1950s, is widely known as the "split brain"—a phenomenon so at odds with common sense that, even within the culture of science, it has defied integration into our thoughts.

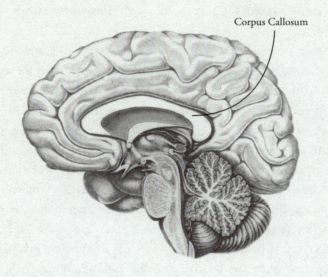

Corpus Callosum

The human brain is divided at the level of the cerebrum (everything above the brain stem) into right and left hemispheres. The reason for this is still unclear, but it does not seem altogether strange that the left-right symmetry of our bodies would be reflected in our central nervous system. This structure turns out to have surprising consequences.

The right and left hemispheres of all vertebrate brains are connected by several nerve tracts called *commissures*, the function of which, we now know, is to pass information back and forth between them. The main commissure in the brains of placental mammals like ourselves is the *corpus callosum*, the fibers of which link similar regions of the cortex across the hemispheres. The evolutionary history of this structure is still a matter of dispute, but in human beings it represents a larger system of connectivity than the sum of all the fibers linking the cortex to the rest of the nervous system.[29] As we are about to see, the unity of every human mind depends on the normal functioning of these connections. Without them, our brains—and minds—are divided.

Certain people have had their forebrain commissures surgically severed. This is generally undertaken as a treatment for severe epilepsy, though other surgeries occasionally require that some of these fibers be cut. As a treatment for epilepsy, patients usually receive a *callosotomy*, a procedure whereby most or all of the corpus callosum is severed to prevent local storms of unregulated activity from spreading throughout the brain and producing a seizure.[30]

The split brain was brought to the world's attention half a century ago by Roger W. Sperry and colleagues.[31] Sperry was awarded a Nobel Prize in 1981 for this work, which inspired a literature that now spans neuroscience, psychology, linguistics, psychiatry, and philosophy. Before Sperry began his research, it appeared that dividing the brains of these patients simply mitigated their seizures (which was, after all, the point) without producing any changes in their behavior. This seemed to lend credence to the ancient notion that the corpus callosum does nothing more than hold the two hemispheres of the brain together.

Once patients recover from this surgery, they generally appear quite normal, even on neurological exam.[32] Under the experimental conditions that Sperry and his colleagues devised, however—first in cats and monkeys,[33] and then in humans[34]—two principal findings emerged. First, the left and right hemispheres of the brain display a high degree of *functional specialization*. This discovery was not entirely new, because it had been known for at least a century that damage to the left hemisphere could impair the use of language. But the split-brain procedure allowed scientists to test each hemisphere independently on a variety of tasks, revealing a range of segregated abilities. The second finding was that when the forebrain commissures are cut, the hemispheres display an altogether astonishing *functional independence*, including separate memories, learning processes, behavioral intentions, and—it seems all but certain—centers of conscious experience.

The independence of the hemispheres in a split-brain patient comes about because most nerve tracts running to and from the cortex are segregated, left and right. Everything that falls in the left visual field of each eye, for instance, is projected to the right hemisphere of the brain, and everything in the right visual field is projected to the left hemisphere. The same pattern holds for both sensation and fine motor control in our extremities. Thus, each hemisphere relies on intact commissures to receive information from its own side of the world. While it can rarely speak, because speech is usually confined to the left hemisphere, the right hemisphere can respond to questions by pointing to written words and objects with the left hand.

The classic demonstration of hemispheric independence in a split-brain patient runs as follows: Show the right hemisphere a word—*egg*, say—by briefly flashing it in the left half of the visual field, and the subject (speaking from his language-dominant left hemisphere) will claim to have seen nothing at all. Ask him to reach behind a partition and select with his left hand (which is predominantly controlled by the right hemisphere) the thing that he "did not see," and he will succeed in picking out an egg from among a multitude of objects. Ask him to name the item he now holds in his left hand without allowing the left hemisphere to get a look at it, and he will be unable to reply. If shown the egg and asked why he selected it from among the available materials, he will probably confabulate an answer (again, with his language-dominant left hemisphere), saying something like "Oh, I picked it because I had eggs for breakfast yesterday." This is a peculiar state of affairs.

When the lateralization of inputs to the brain is exploited in this way, it becomes difficult to say that the person whose brain has been split is a single subject, for everything about his behavior suggests that a silent intelligence lurks in his right hemisphere, about which the articulate left hemisphere knows nothing. The duality of mind is further demonstrated by the fact that these patients can simultaneously perform separate manual tasks. For instance, a person whose brain is functioning normally will find it impossible to draw incompatible figures simultaneously with the right and left hands; divided brains accomplish this task easily, like two artists working in parallel. In the acute phase after surgery, patients' left and right hands sometimes engage in a tug-of-war over an object or sabotage each other's work. The left hemisphere can speak about its condition and may even understand the anatomical details of the procedure that has brought it about, yet it remains remarkably naïve about the experience of its neighbor on the right. Even many years after surgery, the left hemispheres of these subjects express surprise or irritation when their right hemispheres respond to an experimenter's instructions.[35] To ask the left hemisphere what it is like to not know what the right hemisphere is thinking is rather like asking a normal subject what it is like to not know what another person is thinking: He simply *does not know* what the other person is thinking (or even, perhaps, that he or she exists).

What is most startling about the split-brain phenomenon is that we have every reason to believe that the isolated right hemisphere is independently conscious. It is true that some scientists and philosophers have resisted this conclusion,[36] but none have done so credibly. If complex language were necessary for consciousness, then all nonhuman animals and human infants would be devoid of consciousness in principle. If those whose left hemispheres have been surgically removed are still believed to be conscious—and they are—how could the mere presence of a functioning left hemisphere rob the right one of its subjectivity in the case of a split-brain patient?[37]

The consciousness of the right hemisphere is especially difficult to deny whenever a subject possesses linguistic ability on both sides of the brain, because in such cases the divided hemispheres often express different intentions. In a famous example, a young patient was asked what he wanted to be when he grew up: His left brain replied, "A draftsman," while his right brain used letter cards to spell out "racing driver."[38] In fact, the divided hemispheres sometimes seem to address each other directly, in the form of a verbalized, interhemispheric argument.[39]

In such cases, each hemisphere might well have its own beliefs. Consider what this says about the dogma—widely held under Christianity and Islam—that a person's salvation depends upon her believing the right doctrine about God. If a split-brain patient's left hemisphere accepts the divinity of Jesus, but the right doesn't, are we to imagine that she now harbors two immortal souls, one destined for the company of angels and the other for an eternity in hellfire?

The question of whether there is "something that it is like" to be the right hemisphere of a split-brain patient must be answered in the only way that it is ever answered in science: We can merely observe that its behavior and underlying neurology are sufficiently similar to that which we know to be correlated with consciousness in our own case. There is no difficulty in doing this for a normal split-brain patient who retains the use of her left hand. In fact, the consciousness of the disconnected right hemisphere is easier to establish than that of most toddlers. The question of whether the right hemisphere is conscious is really a pseudo-mystery used to bar the door to a great one: the uncanny fact that the human mind can be divided with a knife.

STRUCTURE AND FUNCTION

The right and left hemispheres of our brain show differences in their gross anatomy, many of which are also found in the brains of other animals. In humans, the left hemisphere generally makes a unique contribution to language and to the performance of complex movements. Consequently, damage on this side tends to be accompanied by *aphasia* (impairment of spoken or written language) and *apraxia* (impairment of coordinated movement).

People usually show a right-ear (left-hemisphere) advantage for words, digits, nonsense syllables, Morse code, difficult rhythms, and the ordering of temporal information, whereas they show a left-ear (right-hemisphere) advantage for melodies, musical chords, environmental sounds, and tones of voice. Similar differences have been found for other senses as well. We know, for instance, that the right hand (sensation from which projects almost entirely to the left hemisphere) is better able to discriminate the order of stimuli, while the left hand is more sensitive to their spatial characteristics.

However, the right hemisphere is dominant for many higher cognitive abilities, both in normal brains and in those that have been surgically divided. It tends to have an advantage when reading faces, intuiting geometrical principles and spatial relationships, perceiving wholes from a collection of parts, and judging musical chords.[40] The right hemisphere is also better at displaying emotion (with the left side of the face) and at detecting emotions in others.[41] Interestingly, this obliges us to view one another's least expressive side of the face (the right) with our most emotionally astute hemisphere (the right), and vice versa. Psychopaths generally do not show this right-hemisphere advantage for the perception of emotion; perhaps this is one reason why they are bad at detecting emotional distress in others.[42]

Most evidence suggests that the two hemispheres differ in temperament, and it now seems uncontroversial to say that they can make different (and even opposing) contributions to a person's emotional life.[43] In a divided brain, the hemispheres are unlikely to perceive self and world in the same way, nor are they likely to feel the same about them.

Much of what makes us human is generally accomplished by the right side of the brain. Consequently, we have every reason to believe that the disconnected right hemisphere is independently conscious and that the divided brain harbors two distinct points of view. This fact poses an insurmountable problem for the notion that each of us has a single, indivisible self—much less an immortal soul. The idea of a soul arises from the feeling that our subjectivity has a unity, simplicity, and integrity that must somehow transcend the biochemical wheelworks of the body. But the split-brain phenomenon proves that our subjectivity can quite literally be sliced in two. (This is why Sir John Eccles, a neuroscientist and a committed Christian, declared, against all evidence, that the right hemisphere of the divided brain must be unconscious.) This fact has interesting ethical repercussions. For instance, the biologist Lee Silver wonders what we should do if a person with a split brain wanted to have her right hemisphere removed because she could no longer endure the conflict with her "other self." Would this be a therapeutic intervention or a murder? However, the most important implications are for our view of consciousness: It is divisible—and, therefore, more fundamental than any apparent self.

Imagine undergoing a complete callosotomy. Like most such surgeries, you could be kept awake, because there are no pain receptors in the brain. There is also no reason to think that you would lose consciousness during the procedure, because a person can have an entire hemisphere removed (*hemispherectomy*) without loss of consciousness.[44] Nor would you suffer a lapse in memory. After surgery, you would tend to speak in a way characteristic of *alexithymia* (the inability to describe your emotional life), and you might also demonstrate an inappropriate degree of politeness.[45] Whether or not you had occasion to notice these changes in yourself, it seems all but certain that you would retain your sense of being a "self" throughout the experience.

Given that each hemisphere in your divided brain would have its own point of view, whereas now you appear to have only one, it is natural to wonder which side of the longitudinal fissure "you" would find yourself on once the corpus callosum was cut. Would you land on the right or on the left? It is hard to resist the uncanny demands of arithmetic here. Assuming that you were not simply extinguished and replaced by two new subjects—which seems ruled out by the fact that you would probably remain conscious throughout the procedure and retain your memories—it is tempting to conclude that your subjectivity must collapse to a single hemisphere. Once the surgery was over, it would be obvious that *you* can't be on both sides of the great divide.

Perhaps it is reasonable to believe that you would find yourself in the left hemisphere, retaining the reins of speech, since speech and discursive thinking do much to define your experience in the present. But consider some of the other cognitive abilities you now consciously enjoy, which we know are governed primarily by your right hemisphere. Who, for instance, would greet your loved ones with your left hand and effortlessly recognize their faces, their facial expressions, and their tones of voice?

I think this riddle admits of a rather straightforward solution. Consciousness—whatever its relation to neural events—is divisible. And just as it isn't shared between the brains of separate individuals, it need not be shared between the hemispheres of a single brain once the structures that facilitate such sharing have been cut. If some way of linking two brains with an artificial commissure were ever devised, we should expect that what had been two distinct persons would be unified in the only sense that *consciousness* is ever unified, as a single point of view, and unified in the only sense that *minds* are ever unified, by virtue of common contents and functional abilities.

The experience of dreaming is instructive here. Each night, we lie down to sleep, only to be stolen from our beds and plunged into a realm where our personal histories and the laws of nature no longer apply. Generally, we do not retain enough of a purchase on reality to even notice that anything out of the ordinary has happened. The most astonishing quality of dreams is surely our *lack* of astonishment when they arise. The sleeping brain seems to have no expectation of continuity from one moment to the next. (This is probably owing to the diminished activity in the frontal lobes that occurs during REM sleep.) Thus, sweeping changes in our experience do not, in principle, detract from the unity of consciousness. Left to its own devices, consciousness seems happy to just experience one thing after the next.

If my brain harbors only one conscious point of view—if all that is remembered, intended, and perceived is known by a single "subject"—then I enjoy unity of mind. The evidence is overwhelming, however, that such unity, if it ever exists in a human being, depends upon some humble tracts of white matter crossing the midline of the brain.

ARE OUR MINDS ALREADY SPLIT?

Roger Sperry and his colleagues demonstrated in the 1950s that the corpus callosum cannot facilitate a complete transfer of learning between the cerebral hemispheres.[46] After cutting the optic chiasma in cats (and thereby confining the inputs from each eye to a single hemisphere), they discovered that only simple learning acquired through one eye could transfer to the other side of the brain. Given the immense amount of information processing that takes place in each hemisphere, it seems certain that even a normal human brain will be functionally split to one or another degree. Two hundred million nerve fibers seem insufficient to integrate the simultaneous activity of 20 billion neurons in the cerebral cortex, each of which makes hundreds or thousands (sometimes tens of thousands) of connections to its neighbors.[47] Given this partitioning of information, how can our brains not harbor multiple centers of consciousness even now?

The philosopher Roland Puccetti once observed that the existence of separate spheres of consciousness in the normal brain would explain one of the most perplexing features of split-brain research: Why is it that the right hemisphere is generally willing to bear silent witness to the errors and confabulations of the left? Could it be that the right hemisphere is used to it?

> An answer consistent with the hypothesis of mental duality in the normal human brain suggests itself. The nonspeaking hemisphere has known the true state of affairs from a very tender age. It has known this because beginning at age two or three it heard speech emanating from the common body that, as language development on the left proceeded, became too complex grammatically and syntactically for it to believe it was generating; the same, of course, for what it observed the preferred hand writing down in school through the years. Postsurgically, little has changed for the mute hemisphere (other than loss of sensory information about the ipsilateral half of bodily space). . . . Being inured to this status of cerebral helot, it goes along. Thankless cooperation can become a way of life.[48]

Take a moment to absorb how bizarre this possibility is. The point of view from which you are consciously reading these words may not be the only *conscious* point of view to be found in your brain. It is one thing to say that you are unaware of a vast amount of activity in your brain. It is quite another to say that some of this activity is aware of itself and is watching your every move.

There must be a reason why the structural integrity of the corpus callosum creates a functional unity of mind (insofar as it does), and perhaps it is only the division of the corpus callosum that makes for separated regions of consciousness in the human brain. But whatever the final lesson of the split brain is, it thoroughly violates our commonsense intuitions about the nature of our subjectivity.

A person's experience of the world, while apparently unified in a normal brain, can be physically divided. The problem this poses for the study of consciousness may be insurmountable. If I were to interrogate my brain with the help of a colleague—one who was willing to expose my cortex and begin probing with a microelectrode—neither of us would know what to make of a region that failed to influence the contents of "my" consciousness. The split-brain phenomenon suggests that all that I would be able to say is whether I (as perhaps only one among many centers of consciousness to be found in my brain) felt anything when my friend applied the current. Feeling nothing, I wouldn't know whether the neurons in question constituted a region of consciousness in their own right—for the simple reason that I might be just like a split-brain patient given to wonder, with his articulate left hemisphere, whether or not his right hemisphere is conscious. It surely is, and yet no amount of experimental probing on his part will drive the relevant facts into view. As long as we must correlate changes in the brain—or any other physical system—with first-person reports, any physical systems that are functionally mute may nevertheless prove to be conscious, and our attempt to understand the causes of consciousness will fail to take them into account.

All brains—and persons—may be split to one or another degree. Each of us may live, even now, in a fluid state of split and overlapping subjectivity. Whether or not this seems plausible to *you* may not be the point. Another part of your brain may see the matter differently.

CONSCIOUS AND UNCONSCIOUS
PROCESSING IN THE BRAIN

The frontier between conscious and unconscious mental processes has fascinated psychologists and neuroscientists for more than a century. The realization that the unconscious mind must have some cognitive and emotional structure was the foundation of Freud's work and also the stage upon which he erected an impressively unscientific mythology. The connection between conscious thoughts and unconscious processes was also present in the work of William James, whose views on this topic, and on the mind in general, still deserve our attention:

> Suppose we try to recall a forgotten name. The state of our consciousness is peculiar. There is a gap therein; but no mere gap. It is a gap that is intensely active. A sort of wraith of the name is in it, beckoning us in a given direction, making us at moments tingle with the sense of our closeness, and then letting us sink back without the longed-for term. If wrong names are proposed to us, this singularly definite gap acts immediately so as to negate them. They do not fit into its mould. And the gap of one word does not feel like the gap of another, all empty of content as both might seem necessarily to be when described as gaps. . . . The rhythm of a lost word may be there without a sound to clothe it; or the evanescent sense of something which is the initial vowel or consonant may mock us fitfully, without growing more distinct.[49]

In other words, the unconscious mind exists, and our conscious experience gives some indication of its structure. Recent advances in experimental psychology and neuroimaging have allowed us to study the boundary between conscious and unconscious mental processes with increasing precision. We now know that at least two systems in the brain—often referred to as "dual processes"— govern human cognition, emotion, and behavior. One is evolutionarily older, unconscious, and automatic; the other evolved more recently and is both conscious and deliberative. When you find another person annoying, sexually attractive, or inadvertently funny, you are experiencing the percolations of System 1. The heroic efforts you make to conceal these feelings out of politeness are the work of System 2.

Scientists have learned how to target System 1 through the phenomenon of "priming," revealing that complex mental processes lurk beneath the level of conscious awareness.[50] The experimental technique of "backward masking" has been at the center of this research: Human beings can consciously perceive very brief visual stimuli (down to about $\frac{1}{30}$ of a second), but we can no longer see these images if they are immediately followed by a dissimilar pattern (a "mask"). This fact allows for words and pictures to be delivered to the mind subliminally,[51] and these stimuli have subsequent effects on a person's cognition and behavior. For instance, you will be faster to recognize that *ocean* is a word if it follows a related prime, like *wave*, than if it follows an unrelated one, like *hammer*. And emotionally charged terms are more easily recognized than neutral ones (*sex* can be presented more briefly than *car*), which further demonstrates that the meanings of words must be gleaned prior to consciousness. Subliminally promised rewards drive activity in the brain's reward centers,[52] and masked fearful faces and emotional words increase activity in the amygdala.[53] Clearly, we are not aware of all the information that influences our thoughts, feelings, and actions.

Many other findings attest to the importance of our unconscious mental lives. Amnesiacs, who can no longer form conscious memories, can still improve their performance on a wide variety of tasks through practice.[54] For instance, a person can learn to play golf with increasing proficiency, all the while believing that whenever she picks up a club it is for the first time. The acquisition of such motor skills occurs outside of consciousness in normal people as well. Your conscious memories of practicing a musical instrument, driving a car, or tying your shoelaces are neurologically distinct from your *learning* how to do these things and from your *knowing* how to do them now. People with amnesia can even learn new facts and have their ability to recognize names[55] and generate concepts[56] improve in response to prior exposure, without having any memory of acquiring such knowledge. In fact, we are all in this position with respect to most of our semantic knowledge of the world. Do you remember learning the meaning of the word *door*? Probably not. How do you recognize it and bring its meaning to mind? You have no idea. These processes occur outside consciousness.[57]

CONSCIOUSNESS IS WHAT MATTERS

Despite the obvious importance of the unconscious mind, consciousness is what matters to us—not just for the purpose of spiritual practice but in every aspect of our lives. Consciousness is the substance of any experience we can have or hope for, now or in the future. If God spoke to Moses out of a burning bush, the bush would have been a visual percept (whether veridical or not) of which Moses was consciously aware. It should be clear that if a person begins to suffer from intractable pain or depression, if he experiences a continuous ringing in his ears or the consequences of having acquired a bad reputation among his colleagues, these developments are matters of consciousness and its contents, whatever the nature of the unconscious processes that give rise to them.

Consciousness is also what gives our lives a moral dimension. Without consciousness, we would have no cause to wonder how we should behave toward other human beings, nor could we care how we were treated in return. Granted, many moral emotions and intuitions operate unconsciously, but it is because they influence the contents of consciousness that they matter to us. I have argued elsewhere, and at length in *The Moral Landscape*, that we have ethical responsibilities toward other creatures precisely to the degree that our actions can affect their conscious experience for better or worse.[58] We don't have ethical obligations toward rocks (on the assumption that they are not conscious), but we do have such obligations toward any creature that can suffer or be deprived of happiness. Of course, it can be wrong to destroy rocks if they happen to be valuable to other conscious creatures. The Taliban's destruction of the 1,500-year-old standing Buddhas of Bamiyan was wrong not from the perspective of the statues themselves but from that of all the people who cared about them (and the future people who might have cared).

I have never come across a coherent notion of bad or good, right or wrong, desirable or undesirable that did not depend upon some change in the experience of conscious creatures. It is not always easy to nail down what we mean by "good" and "bad"—and their definitions may remain perpetually open to revision—but such judgments seem to require, in every instance, that some difference register at the level of experience. Why would it be wrong to murder a billion human beings? Because so much pain and suffering would result. Why would it be wrong to painlessly kill every man, woman, and child in their sleep? Because of all the possibilities for future happiness that would be foreclosed. If you think such actions are wrong primarily because they would anger God or would lead to your punishment after death, you are still worried about perturbations of consciousness—albeit ones that stand a good chance of being wholly imaginary.

I take it to be axiomatic, therefore, that our notions of meaning, morality, and value presuppose the actuality of consciousness (or its loss) *somewhere*. If anyone has a conception of meaning, morality, and value that has nothing to do with the experience of conscious beings, in this world or in a world to come, I have yet to hear of it. And it would seem that such a conception of value could hold no interest for anyone, by definition, because it would be guaranteed to be outside the experience of every conscious being, now and in the future.

The fact that the universe is illuminated where you stand—that your thoughts and moods and sensations have a qualitative character in this moment—is a mystery, exceeded only by the mystery that there should be something rather than nothing in the first place. Although science may ultimately show us how to truly maximize human well-being, it may still fail to dispel the fundamental mystery of our being itself. That doesn't leave much scope for conventional religious beliefs, but it does offer a deep foundation for a contemplative life. Many truths about ourselves will be discovered in consciousness directly or not discovered at all.

Chapter 3

The Riddle of the Self

I once spent an afternoon on the northwestern shore of the Sea of Galilee, atop the mount where Jesus is believed to have preached his most famous sermon. It was an infernally hot day, and the sanctuary where I sat was crowded with Christian pilgrims from many continents. Some gathered silently in the shade, while others staggered about in the sun, taking photographs.

As I gazed at the surrounding hills, a feeling of peace came over me. It soon grew to a blissful stillness that silenced my thoughts. In an instant, the sense of being a separate self—an "I" or a "me"—vanished. Everything was as it had been—the cloudless sky, the brown hills sloping to an inland sea, the pilgrims clutching their bottles of water—but I no longer felt separate from the scene, peering out at the world from behind my eyes. Only the world remained.

The experience lasted just a few seconds, but it returned many times as I looked out over the land where Jesus is believed to have walked, gathered his apostles, and worked many of his miracles. If I were a Christian, I would undoubtedly have interpreted this experience in Christian terms. I might believe that I had glimpsed the oneness of God or been touched by the Holy Spirit. If I were a Hindu, I might think in terms of Brahman, the eternal Self, of which the world and all individual minds are thought to be a mere modification. If I were a Buddhist, I might talk about the "dharmakaya of emptiness," in which all apparent things manifest as in a dream.

But I am simply someone who is making his best effort to be a rational human being. Consequently, I am very slow to draw metaphysical conclusions from experiences of this sort. And yet, I glimpse what I will call the *intrinsic selflessness of consciousness* every day, whether at a traditional holy site, or at my desk, or while having my teeth cleaned. This is not an accident. I've spent many years practicing meditation, the purpose of which is to cut through the illusion of the self.

My goal in this chapter and the next is to convince you that the conventional sense of self is an illusion—and that spirituality largely consists in realizing this, moment to moment. There are logical and scientific reasons to accept this claim, but recognizing it to be true is not a matter of understanding these reasons. Like many illusions, the sense of self disappears when closely examined, and this is done through the practice of meditation. Once again, I am suggesting an experiment that you must conduct for yourself, in the laboratory of your own mind, by paying attention to your experience in a new way.

The Buddha's famous parable meant to denigrate mere intellectualism seems apropos here:[1] A man is struck in the chest with a poison arrow. A surgeon rushes to his side to begin the work of saving his life, but the man resists these ministrations. He first wants to know the name of the fletcher who fashioned the arrow's shaft, the genus of the wood from which it was cut, the disposition of the man who shot it, the name of the horse upon which he rode, and a thousand other things that have no bearing upon his present suffering or his ultimate survival. The man needs to get his priorities straight. His commitment to *thinking about* the world results from a basic misunderstanding of his predicament. And though we may be only dimly aware of it, we, too, have a problem that will not be solved by acquiring more conceptual knowledge.

Little has changed since the Buddha's time. Many people claim to have no interest at all in spiritual life. Indeed, most scientists and philosophers disdain the subject, for it suggests a neglect of intellectual standards: Bliss, it has been noted, is not conducive to detached observation.[2] And yet, we are all seeking fulfillment while living at the mercy of changing experience. Whatever we acquire in life gets dispersed. Our bodies age. Our relationships fall away. Even the most intense pleasures last only a few moments. And every morning, we are chased out of bed by our thoughts.

In this chapter, I will invoke a variety of concepts that have yet to do much useful work in our study of the natural world, or even of the brain, but do very heavy lifting throughout the course of our lives: concepts such as *self* and *ego* and *I*. Admittedly, these terms appear less than scientific, but we have no new words with which to name, and subsequently study, one of the most striking features of our existence: Most of us feel that our experience of the world refers back to a self—not to our bodies precisely but to a center of consciousness that exists somehow interior to the body, behind the eyes, inside the head. The feeling that we call "I" seems to define our point of view in every moment, and it also provides an anchor for popular beliefs about souls and freedom of will. And yet this feeling, however imperturbable it may appear at present, can be altered, interrupted, or entirely abolished. Such transformations run the gamut from run-of-the-mill psychosis to spiritual epiphany.

What makes me the same person I was five minutes ago, or yesterday, or on my eighteenth birthday? Is it that I remember being those former selves and my memories are (somewhat) accurate? In fact, I've forgotten most of what has happened to me over the course of my life, and my body has been gradually changing all the while. Is it enough to say that I am physically continuous with my former selves because most of the cells in my body are the same as or descended from those that made up the bodies of these younger men?

As we have seen, the split-brain phenomenon puts pressure on the very idea of personal identity. But things can get even worse. In a now famous thought experiment, the philosopher Derek Parfit asks us to imagine a teleportation device that can beam a person from Earth to Mars. Rather than travel for many months on a spaceship, you need only enter a small chamber close to home and push a green button, and all the information in your brain and body will be sent to a similar station on Mars, where you will be reassembled down to the last atom.

Imagine that several of your friends have already traveled to Mars this way and seem none the worse for it. They describe the experience as being one of instantaneous relocation: You push the green button and find yourself standing on Mars—where your most recent memory is of pushing the green button on Earth and wondering if anything would happen.

So you decide to travel to Mars yourself. However, in the process of arranging your trip, you learn a troubling fact about the mechanics of teleportation: It turns out that the technicians wait for a person's replica to be built on Mars before obliterating his original body on Earth. This has the benefit of leaving nothing to chance; if something goes wrong in the replication process, no harm has been done. However, it raises the following concern: While your double is beginning his day on Mars with all your memories, goals, and prejudices intact, you will be standing in the teleportation chamber on Earth, just staring at the green button. Imagine a voice coming over the intercom to congratulate you for arriving safely at your destination; in a few moments, you are told, your Earth body will be smashed to atoms. How would this be any different from simply being killed?

To most readers, this thought experiment will suggest that *psychological* continuity—the mere maintenance of one's memories, beliefs, habits, and other mental traits—is an insufficient basis for personal identity. It's not enough for someone on Mars to be just like you; he must *actually be* you. The man on Mars will share all your memories and will behave exactly as you would have. But he is not *you*—as your continued existence in the teleportation chamber on Earth attests. To the Earth-you awaiting obliteration, teleportation as a means of travel will appear a horrifying sham: You never left Earth and are about to die. Your friends, you now realize, have been repeatedly copied and killed. And yet, the problem with teleportation is somehow not obvious if a person is disassembled before his replica is built. In that case, it is tempting to say that teleportation works and that "he" is really stepping onto the surface of Mars.

One might conclude that personal identity requires *physical* continuity: I am identical to my brain and body, and if they get destroyed, that's the end of me. But Parfit shows that physical continuity matters only because it normally supports psychological continuity. Merely hanging on to one's brain and body cannot be an end in itself. Just consider the unfortunate case of someone with advanced dementia: He is physically but not psychologically continuous with the person he used to be. If he could be given new neurons that would emulate the old ones in his healthy brain—restoring his memories, creativity, sense of humor—this would be far better than keeping his current neurons that are succumbing to neurodegenerative disease. If we grant that the gradual replacement of individual neurons would be compatible with continued consciousness, it seems clear that the maintenance of psychological continuity is what we care about. And it is generally what we mean by a person's "survival" from one moment to the next.

Parfit pushes the concept of personal identity about as far as it can go and resolves the apparent paradox of teleportation by arguing that "identity is not what matters"; rather, we should be concerned only about psychological continuity. However, he also states that psychological continuity cannot take a "branching form" (or at least not for long), as it does when a person is copied on Mars while the original person survives on Earth. Parfit believes that we should view the teleportation case in which a person is destroyed before being replicated as more or less indistinguishable from the normal pattern of personal survival throughout our lives. After all, in what way are you subjectively the same as the person who first picked up this book? In the only way you *can* be: by displaying some degree of psychological continuity with that past self. Viewed in this way, it is difficult to see how teleportation is any different from the mere passage of time. As Parfit says, "I want the person on Mars to be me in a specially intimate way in which no future person will ever be me. . . . What I fear will be missing is *always* missing. . . . *Ordinary survival is about as bad as being destroyed and Replicated.*"[3] Here, Parfit does not mean "bad" in the sense that we should find these truths depressing. He is merely arguing that ordinary survival from moment to moment is no more demonstrative of personal identity than destruction/ replication would be. Parfit's view of the self, which he appears to have arrived at independently through an immensely creative use of thought experiments, is essentially the same as the one found in the teachings of Buddhism: There is no stable self that is carried along from one moment to the next.

I agree with most of what Parfit has to say about personal identity. However, because his view is purely the product of logical argument, it can seem uncannily detached from the reality of our lives. Although experience in meditation may not immediately resolve the teleportation paradox or make it clear why one should care about one's own future experience any more than that of a stranger, it can make these philosophical problems easier to think about.

When talking about psychological continuity, we are talking about consciousness and its contents—the persistence of autobiographical memories in particular. Everything that is personal, everything that differentiates my consciousness from that of another human being, relates to the *contents* of consciousness. Memories, perceptions, attitudes, desires—these are appearances in consciousness. If "my" consciousness were suddenly filled with the contents of "your" life—if I awoke this morning with your memories, hopes, fears, sensory impressions, and relationships— I would no longer be me. I would be the same as your clone in the teleportation case.

My consciousness is "mine" only because the particularities of my life are illuminated as and where they arise. For instance, I currently have an annoying pain in my neck, the result of a martial arts injury. Why is this "my" pain? Why am I the only one who is directly aware of it? These questions are a symptom of confusion. There is no "I" who is aware of the pain. The pain is simply arising in consciousness in the only place it *can* arise: at the conjunction of this brain and this neck. Where else could this particular pain be felt? If I were cloned through teleportation, an identical pain might be felt in an identical neck on Mars. But *this* pain would still be right here in *this* neck.

Whatever its relation to the physical world, consciousness is the context in which the objects of experience appear—the sight of this book, the sound of traffic, the sensation of your back against a chair. There is nowhere else for them *to* appear—for their very appearance is consciousness in action. And anything that is unique to your experience of the world must appear amid the contents of consciousness. We have every reason to believe that these contents depend upon the physical structure of your brain. Duplicate your brain, and you will duplicate "your" contents in another field of consciousness. Divide your brain, and you will segregate those contents in bizarre ways.

We know, from experiments both real and imagined, that psychological continuity is divisible—and can, therefore, be inherited by more than one mind. If my brain were surgically divided by callosotomy tomorrow, this would create at least two independent conscious minds, both of which would be psychologically continuous with the person who is now writing this paragraph. If my linguistic abilities happened to be distributed across both hemispheres, each of these minds might remember having written this sentence. The question of whether *I* would land in the left hemisphere or the right doesn't make sense—being based, as it is, on the illusion that there is a self bobbing on the stream of consciousness like a boat on the water.

But the stream of consciousness can divide and follow both tributaries simultaneously. Should these tributaries converge again, the final current would inherit the "memories" of each. If, after years of living apart, my hemispheres were reunited, their memories of separate existence could, in principle, appear as the combined memory of a single consciousness. There would be no cause to ask where my "self" had been while my brain was divided, because no "I" exists apart from the stream. The moment we see this, the divisibility of the human mind begins to seem less paradoxical. Subjectively speaking, the only thing that actually exists is consciousness and its contents. And the only thing relevant to the question of personal identity is psychological continuity from one moment to the next.

WHAT ARE WE CALLING "I"?

One thing each of us knows for certain is that reality vastly exceeds our awareness of it. I am, for instance, sitting at my desk, drinking coffee. Gravity is holding me in place, and the manner in which this is accomplished eludes us to this day. The integrity of my chair is the result of the electrical bonds between atoms—entities I have never seen but which I know must exist, in some sense, with or without my knowledge. The coffee is dissipating heat at a rate that could be calculated with precision, and the second law of thermodynamics decrees that it is, on balance, losing heat every moment rather than gathering it from the cup or the surrounding air. None of this is evident to me from direct experience, however. Forces of digestion and metabolism are at work within me that are utterly beyond my perception or control. Most of my internal organs may as well not exist for all I know of them directly, and yet I can be reasonably certain that I have them, arranged much as any medical textbook would suggest. The taste of the coffee, my satisfaction at its flavor, the feeling of the warm cup in my hand—while these are immediate facts with which I am acquainted, they reach back into a dark wilderness of facts that I will never come to know. I have neurons firing and forming new connections in my brain every instant, and these events determine the character of my experience. But I know nothing directly about the electrochemical activity of my brain—and yet this soggy miracle of computation appears to be working for the moment and generating a vision of a world.

The more I persist in this line of thought, the clearer it becomes that I perceive scarcely a scintilla of all that exists to be known. I can, for instance, reach for my cup of coffee or set it down, seemingly as I please. These are intentional actions, and *I* perform them. But if I look for what underlies these movements—motor neurons, muscle fibers, neurotransmitters—I can't feel or see a thing. And *how* do I initiate this behavior? I haven't a clue. In what sense, then, do *I* initiate it? That is difficult to say. The feeling that I intended to do what I just did seems to be only that: a *feeling* of some internal signature, perhaps the result of my brain's having formed a predictive model of its ensuing actions. It may not be best classified as a feeling, but surely it is *something*. Otherwise, how could I note the difference between voluntary and involuntary behavior? Without this impression of agency, I would feel that my actions were automatic or otherwise beyond my control.

One question immediately presents itself: Where am I that I have such a poor view of things? And what sort of thing am I that *both* my outside and my inside are so obscure? And outside and inside of *what*? My *skin*? Am I identical to my skin? If not—and the answer is clearly no—why should the frontier between my outside and my inside be drawn at the skin? If not at the skin, then where does the outside of me stop and the inside of me begin? At my skull? Am I my skull? Am I *inside* my skull? Let's say yes for the moment, because we are quickly running out of places to look for me. Where inside my skull might I be? And if I'm up there in my head, how is the rest of me *me* (let alone the *inside* of me)?

The pronoun *I* is the name that most of us put to the sense that we are the thinkers of our thoughts and the experiencers of our experience. It is the sense that we have of possessing (rather than of merely being) a continuum of experience. We will see, however, that this feeling is not a necessary property of the mind. And the fact that people report losing their sense of self to one or another degree suggests that the experience of being a self can be selectively interfered with.

Obviously, there is something in our experience that we are calling "I," apart from the sheer fact that we are conscious; otherwise, we would never describe our subjectivity in the way we do, and a person would have no basis for feeling that she had *lost* her sense of self, whatever the circumstances. Nevertheless, it is extremely difficult to pinpoint just what it is we take ourselves to be. Many philosophers have noticed this problem, but few in the West have understood that the failure to locate the self can produce more than mere confusion.[4] I suspect that this difference between Eastern and Western philosophy has something to do with the influence of Abrahamic religion and its doctrine of the soul. Christianity, in particular, presents impressive obstacles to thinking intelligently about the nature of the human mind, asserting, as it does, the real existence of individual souls who are subject to the eternal judgment of God.

What does it mean to say that the self cannot be found or that it is illusory? It is not to say that *people* are illusory. I see no reason to doubt that each of us exists or that the ongoing history of our personhood can be conventionally described as the history of our "selves." But the self in this more global, biographical sense undergoes sweeping changes over the course of a lifetime.

While you are in many ways physically and psychologically continuous with the person you were at age seven, you are not the same. Your life has surely been punctuated by transitions that significantly changed you: marriage, divorce, college, military service, parenthood, bereavement, serious illness, fame, exposure to other cultures, imprisonment, professional success, loss of a job, religious conversion. Each of us knows what it is like to develop new capacities, understandings, opinions, and tastes over the course of time. It is convenient to ascribe these changes to the self. That is not the self I am talking about.

The self that does not survive scrutiny is the *subject* of experience in each present moment—the feeling of being a thinker of thoughts *inside* one's head, the sense of being an owner or inhabitant of a physical body, which this false self seems to appropriate as a kind of vehicle. Even if you don't believe such a homunculus exists—perhaps because you believe, on the basis of science, that you are identical to your body and brain rather than a ghostly resident therein—you almost certainly *feel* like an internal self in almost every waking moment. And yet, however one looks for it, this self is nowhere to be found. It cannot be seen amid the particulars of experience, and it cannot be seen when experience itself is viewed as a totality. However, its *absence* can be found—and when it is, the feeling of being a self disappears.

CONSCIOUSNESS WITHOUT SELF

This is an empirical claim: Look closely enough at your own mind in the present moment, and you will discover that the self is an illusion. The problem with a claim of this kind, however, is that one can't borrow another person's contemplative tools to test it. To see how the feeling of "I" is a product of thought—indeed, to even appreciate how distracted by thought you tend to be in the first place—you have to build your own contemplative tools. Unfortunately, this leads many people to dismiss the project out of hand: They look inside, notice nothing of interest, and conclude that introspection is a dead end. But just imagine where astronomy would be if, centuries after Galileo, a person were still obliged to build his own telescope before he could even judge whether astronomy was a legitimate field of inquiry. It wouldn't make the sky any less worthy of investigation, but astronomy's development as a science would become immensely more difficult.

A few pharmacological shortcuts exist—and I discuss some of them in a later chapter—but generally speaking, we must build our own telescopes to judge the empirical claims of contemplatives. Judging their metaphysical claims is another matter; many of them can be dismissed as bad science or bad philosophy after merely thinking about them. But to determine whether certain experiences are possible—and if possible, desirable—and to see how these states of mind relate to the conventional sense of self, we have to be able to use our attention in the requisite ways. Primarily, that means learning to recognize thoughts as *thoughts*— as transient appearances in consciousness—and to no longer be distracted by them, if only for short periods of time. This may sound simple enough, but actually accomplishing it can take a lot of work. Unfortunately, it is not work that the Western intellectual tradition knows much about.

LOST IN THOUGHT

When we see a person walking down the street talking to himself, we generally assume that he is mentally ill (provided he is not wearing a headset of some kind). But we all talk to ourselves *constantly*—most of us merely have the good sense to keep our mouths shut. We rehearse past conversations—thinking about what we said, what we didn't say, what we should have said. We anticipate the future, producing a ceaseless string of words and images that fill us with hope or fear. We tell ourselves the story of the present, as though some blind person were inside our heads who required continuous narration to know what is happening: "Wow, nice desk. I wonder what kind of wood that is. Oh, but it has no drawers. They didn't put drawers in this thing? How can you have a desk without at least one drawer?" Who are we talking to? No one else is there. And we seem to imagine that if we just keep this inner monologue to ourselves, it is perfectly compatible with mental health. Perhaps it isn't.

As I was working to finish this book, we experienced a series of plumbing leaks in our house. The first appeared in the ceiling of a storage room. We considered ourselves genuinely lucky to have found it, because this was a room that we might have gone months without entering. A plumber arrived within a few hours, cut the drywall, and fixed the leak. A plasterer came the next day, repaired the ceiling, and painted it. This sort of thing happens eventually in every home, I told myself, and my prevailing feeling was of gratitude. Civilization is a wonderful thing.

Then a similar leak appeared in an adjacent room a few days later. Contact information for both the plumber and the plasterer was at my fingertips. Now I felt only annoyance and foreboding.

A month later, the horror movie began in earnest: A pipe burst, flooding six hundred square feet of ceiling. This time the repair took weeks and created an immense amount of dust; two cleaning crews were required to deal with the aftermath—vacuuming hundreds of books, drying and shampooing the carpet, and so forth. Throughout all this we were forced to live without heat, for otherwise the dust from the repair would have been sucked into the vents, and we would have been breathing it in every room of the house. Eventually, however, the problem was fixed. We would have no more leaks.

And then, last night, scarcely one month after the previous repair, we heard the familiar sound of water falling onto carpet. The moment I heard the first drops, I was transformed into a hapless, uncomprehending, enraged man racing down a staircase. I'm sure I would have comported myself with greater dignity had I come upon the scene of a murder. A glance at the ballooning ceiling told me everything I needed to know about the weeks ahead: Our home would be a construction site once again.

Of course, a house is a physical object beholden to the laws of nature—and it won't fix itself. From the moment my wife and I grabbed buckets and salad bowls to catch the falling water, we were responding to the ineluctable tug of physical reality. But my suffering was entirely the product of my thoughts. Whatever the needs of the moment, I had a choice: I could do what was required calmly, patiently, and attentively, or do it in a state of panic. Every moment of the day—indeed, every moment throughout one's life—offers an opportunity to be relaxed and responsive or to suffer unnecessarily.

We can address mental suffering of this kind on at least two levels. We can use thoughts themselves as an antidote, or we can stand free of thought altogether. The first technique requires no experience with meditation, and it can work wonders if one develops the appropriate habits of mind. Many people do it quite naturally; it's called "looking on the bright side."

For instance, as I was beginning to rage like King Lear in the storm, my wife suggested that we should be thankful that it was fresh water pouring through our ceiling and not sewage. I found the thought immediately arresting: I could feel in my bones how much better it was to be mopping up water at that moment than to be ankle deep in the alternative. What a relief! I often use thoughts of this kind as levers to pry my mind loose from whatever rut it has found on the landscape of unnecessary suffering. If I had been watching sewage spill through our ceiling, how much would I have paid merely to transform it into fresh water? A lot.

I am not advocating that we be irrationally detached from the reality of our lives. If a problem needs fixing, we should fix it. But how miserable must we be while doing good and necessary things? And if, like many people, you tend to be vaguely unhappy much of the time, it can be very helpful to manufacture a feeling of gratitude by simply contemplating all the terrible things that have *not* happened to you, or to think of how many people would consider their prayers answered if they could only live as you are now. The mere fact that you have the leisure to read this book puts you in very rarefied company. Many people on earth at this moment can't even *imagine* the freedom that you currently take for granted.

In fact, the effects of consciously practicing gratitude have been studied: When compared to merely thinking about significant life events, contemplating daily hassles, or comparing oneself favorably to others, thinking about what one is grateful for increases one's feelings of well-being, motivation, and positive outlook toward the future.[5]

One does not need to know anything about meditation to notice how thinking governs one's mental state. This morning, for instance, I awoke in a state of carefree happiness. And then I remembered the leak. . . . Most readers will be familiar with this experience: Something bad has happened in your life—a person has died, a relationship has ended, you have lost your job—but there is a brief interval after awakening before memory imposes its stranglehold. It often takes a moment or two for one's reasons for being unhappy to come online. Having spent years observing my mind in meditation, I find such sudden transitions from happiness to suffering both fascinating and rather funny—and merely witnessing them goes a long way toward restoring my equanimity. My mind begins to seem like a video game: I can either play it intelligently, learning more in each round, or I can be killed in the same spot by the same monster, again and again.

Once, while staying in an especially depressing hotel in Kathmandu, I was awakened in the middle of the night by the feeling of a claw scratching my foot. I sat up in terror, convinced that there was a rat in my bed. I had recently learned that the lepers I had seen throughout my travels in Asia lost their fingers and toes not to the disease itself but because they no longer felt pain. This resulted in burns and other injuries. Even worse, rats often ate their extremities while they slept.

However, the darkness of my room was perfectly still. It had been only a dream. And as suddenly as it had come, the feeling of terror subsided. My mind and body were now flooded with relief. "What a strange dream," I thought. "I actually *felt* claws on my skin—but nothing was there. The mind is so amazing"—and then came the unmistakable sound of something scurrying toward me beneath the sheets.

I bounded from the bed with the agility of a Chinese acrobat. After a few interminable moments spent groping in the darkness of an unfamiliar room, I turned on the lights, and all was silent once again. As I stared at the tangle of blankets on the bed, I genuinely hoped that I had lost my sanity and not, in fact, my privacy. I flung back the covers—and there, in the middle of the mattress, sat a large brown rat. The creature eyed me with a sickening frank-

ness and intensity; it appeared to be *standing its ground*, no doubt
ruing the loss of such an ample source of protein. I feigned an
attack of my own, lunging and shrieking—half ape, half cartoon
housewife—and the beast raced across the sheets, sprung to the
floor, and disappeared behind the dresser.[6]

In the span of a few seconds, my mind had traversed the extremes of human emotion, swinging from terror to exquisite relief
and back to terror—entirely on the wings of thought:

There's a rat in my bed!
Oh, it was only a dream . . .
Rat!

Again, I'm not saying that one's thoughts about reality are all
that matter. I would be the first to admit that it is generally a good
idea to keep rats out of one's bed. But it can be liberating to see
how thoughts pull the levers of emotion—and how negative emotions in turn set the stage for patterns of thinking that keep them
active and coloring one's mind. Seeing this process clearly can
mean the difference between being angry, depressed, or fearful for
a few moments and being so for days, weeks, and months on end.

Breaking the Spell of Negative Emotions

Most of us let our negative emotions persist longer than
is necessary. Becoming suddenly angry, we tend to stay
angry—and this requires that we actively produce the
feeling of anger. We do this by thinking about our reasons for being angry—recalling an insult, rehearsing what
we should have said to our malefactor, and so forth—and

yet we tend not to notice the mechanics of this process. Without continually resurrecting the feeling of anger, it is impossible to stay angry for more than a few moments.

While I can't promise that meditation will keep you from ever again becoming angry, you can learn not to stay angry for very long. And when talking about the consequences of anger, the difference between moments and hours—or days—is impossible to exaggerate.

Even without knowing how to meditate, most people have experienced having their negative states of mind suddenly interrupted. Imagine, for instance, that someone has made you very angry—and just as this mental state seems to have fully taken possession of your mind, you receive an important phone call that requires you to put on your best social face. Most people know what it's like to suddenly drop their negative state of mind and begin functioning in another mode. Of course, most then helplessly grow entangled with their negative emotions again at the next opportunity.

Become sensitive to these interruptions in the continuity of your mental states. You are depressed, say, but are suddenly moved to laughter by something you read. You are bored and impatient while sitting in traffic, but then are cheered by a phone call from a close friend. These are natural experiments in shifting mood. Notice that suddenly paying attention to something else—something that no longer supports your current emotion—allows for a new state of mind. Observe how quickly the clouds can part. These are genuine glimpses of freedom.

The truth, however, is that you need not wait for some pleasant distraction to shift your mood. You can simply pay close attention to negative feelings themselves, without judgment or resistance. What is anger? Where do you feel it in your body? How is it arising in each moment? And what is it that is aware of the feeling itself? Investigating in this way, with mindfulness, you can discover that negative states of mind vanish all by themselves.

Thinking is indispensable to us. It is essential for belief formation, planning, explicit learning, moral reasoning, and many other capacities that make us human. Thinking is the basis of every social relationship and cultural institution we have. It is also the foundation of science. But our habitual identification with thought—that is, our failure to recognize thoughts *as thoughts*, as appearances in consciousness—is a primary source of human suffering. It also gives rise to the illusion that a separate self is living inside one's head.

See if you can stop thinking for the next sixty seconds. You can notice your breath, or listen to the birds, but do not let your attention be carried away by thought, *any thought*, even for an instant. Put down this book, and give it a try.

Some of you will be so distracted by thought as to imagine that you succeeded. In fact, beginning meditators often think that they are able to concentrate on a single object, such as the breath, for minutes at a time, only to report after days or weeks of intensive practice that their attention is now carried away by thought every few seconds. This is actually progress. It takes a certain degree of concentration to even notice how distracted you are. Even if your life depended on it, you could not spend a full minute free of thought.

This is a remarkable fact about the human mind. We are capable of astonishing feats of understanding and creativity. We can endure almost any torment. But it is not within our power to simply stop talking to ourselves, whatever the stakes. It's not even in our power to recognize each thought as it arises in consciousness without getting distracted every few seconds by one of them. Without significant training in meditation, remaining aware—of *anything*—for a full minute is just not in the cards.

We spend our lives lost in thought. The question is, what should we make of this fact? In the West, the answer has been "Not much." In the East, especially in contemplative traditions like those of Buddhism, being distracted by thought is understood to be the very wellspring of human suffering.

From the contemplative point of view, being lost in thoughts of any kind, pleasant or unpleasant, is analogous to being asleep and dreaming. It's a mode of not knowing what is actually happening in the present moment. It is essentially a form of psychosis. Thoughts themselves are not a problem, but being identified with thought is. Taking oneself to be the thinker of one's thoughts—that is, not recognizing the present thought to be a transitory appearance in consciousness—is a delusion that produces nearly every species of human conflict and unhappiness. It doesn't matter if your mind is wandering over current problems in set theory or cancer research; if you are thinking without knowing you are thinking, you are confused about who and what you are.

The practice of meditation is a method of breaking the spell of thought. However, in the beginning, you are unlikely to understand just how transformative this shift in attention can be. You will spend most of your time *trying* to meditate or imagining that you *are* meditating (whether by focusing on your breathing or anything else) and failing for minutes or hours at a stretch. The first sign of progress will be noticing how distracted you are. But if you persist in your practice, you will eventually get a taste of real concentration and begin to see thoughts themselves as mere appearances arising in a wider field of consciousness.

The eighth-century Buddhist adept Vimalamitra described three stages of mastery in meditation and how thinking appears in each. The first is like meeting a person you already know; you simply recognize each thought as it arises in consciousness, without confusion. The second is like a snake tied in a knot; each thought, whatever its content, simply unravels on its own. In the third, thoughts become like thieves entering an empty house; even the possibility of being distracted has disappeared.[7]

Long before reaching this kind of stability in meditation, however, one can discover that the sense of self—the sense that there is a thinker behind one's thoughts, an experiencer amid the flow of experience—is an illusion. The feeling that we call "I" is itself the product of thought. Having an *ego* is what it feels like to be thinking without knowing that you are thinking.

Consider the following train of thought (a version of which may have already passed through your mind):

> What is Harris going on about? I know I'm thinking. I'm thinking right now. What's the big deal? I'm thinking, and I know it. How is this a problem? How am I confused? I can think about anything I want—watch, I'll picture the Eiffel Tower in my mind's eye right now. There it is. I did it. In what sense am I not the thinker of these thoughts?

Thus is the knot of self tied. It isn't enough to know, in the abstract, that thoughts continually arise or that one is thinking at this moment, for such knowledge is itself mediated by thoughts that are arising unrecognized. It is the identification with these thoughts—that is, the failure to recognize them as they spontaneously appear in consciousness—that produces the feeling of "I." One must be able to pay attention closely enough to glimpse what consciousness is like *between* thoughts—that is, prior to the arising of the next one. *Consciousness does not feel like a self.* Once one realizes this, the status of thoughts themselves, as transient expressions of consciousness, can be understood.

What are we conscious of? We are conscious of the world; we are conscious of our bodies in the world; and we also imagine that we are conscious of our selves within our bodies. After all, most of us don't feel merely identical to our bodies. We seem to be riding around *inside* our bodies. We feel like inner subjects that can use the body as a kind of object. This last impression is an illusion that can be dispelled.

The selflessness of consciousness is in plain view in every present moment—and yet, it remains difficult to see. This is not a paradox. Many things in our experience are right on the surface, but they require some training or technique to observe. Consider the optic blind spot: The optic nerve passes through the retina of each eye, creating a small region in each visual field where we are effectively blind. Many of us learned as children to perceive the subjective consequences of this less-than-ideal anatomy by drawing a small circle on a piece of paper, closing one eye, and then moving the paper into a position where the circle became invisible. No doubt most people in human history have been totally unaware of the optic blind spot. Even those of us who know about it go for decades without noticing it. And yet, it is always there, right on the surface of experience.

The absence of the self is also there to be noticed. As with the optic blind spot, the evidence is not far away or deep within; rather, it is almost too close to be observed. For most people, experiencing the intrinsic selflessness of consciousness requires considerable training. It is, however, possible to notice that consciousness—that in you which is aware of your experience in this moment—does not feel like a self. It does not feel like "I." What you are calling "I" is itself a feeling that arises among the contents of consciousness. Consciousness is prior to it, a mere witness of it, and, therefore, free of it in principle.

THE CHALLENGE OF STUDYING THE SELF

Many scientists use the term *self* to refer to the totality of our inner lives. I have attended whole conferences on the self and read books ostensibly devoted to this topic without seeing the feeling we call "I" even mentioned. The self that I am discussing throughout this book—the illusory, albeit reliable, source of so much suffering and confusion—is the feeling that there is an inner subject, behind our eyes, thinking our thoughts and experiencing our experience.

We must distinguish between the self and the myriad mental states—self-recognition, volition, memory, bodily awareness—with which it can be associated. To appreciate the difference, consider the (semi-fictional) condition of a person suffering from global retrograde amnesia (sometimes called "soap opera" amnesia, wherein a person has entirely forgotten his past): If asked how he came to be this way, he might say, "I don't remember anything." This is overstating the case, for he must remember a thing or two (the English language, for instance) to even make such a statement. But there is no reason to think that he is misusing the personal pronoun *I*. His "I" seems to have survived the loss of his declarative memories as fully as his body has. If we asked him, "Where is your body?" he might say, "It's here. This is it." If we questioned him further, asking, "And where are you? Where is your self?" he would probably say something like "What do you mean? I'm here too. I just don't know who I am." Strange as this conversation would be, there seems little doubt that our protagonist would feel as much like a self as we do. Only his memories are missing. He, as the subject of his experience, remains to worry over their absence.

Of course, as a *person*, this man is no longer himself. He doesn't remember the names or the faces of his closest friends. He may not know which foods he likes. His private fears and professional goals have disappeared without a trace. We may say that he is scarcely a person at all—but he is a *self* all the same, and one that is suffering a bewildering dissociation from both past and future.

Or consider the condition of a person who is having an "out-of-body experience" (OBE). The sense of leaving one's body is a staple of mystical literature and has been reported across many cultures. It is often associated with epilepsy, migraine, sleep paralysis, and, as we will see in chapter 5, the "near-death experience." It may occur in as much as 10 percent of the population. During an OBE, the subject feels that she has physically left her body—and this often includes a sense that she can see her own body in full, as though from a point outside her head. A brain area called the *temporal-parietal junction*—a region known to be involved in sensory integration and body representation—seems to be responsible for this effect. Whether or not a person's consciousness can really be displaced is irrelevant; the point is that it can *seem* to be, and this fact draws yet another boundary between the self and the rest of our personhood. It is possible to experience oneself as (apparently) outside a body.

The self, as the implied hub of cognition, perception, emotion, and behavior, can remain stable across even wholesale changes in the contents of consciousness (unless the feeling of self disappears). This is not surprising, because the self is the very thing to which these contents seem to refer: not the body or mind per se but the point of view from which both body and mind seem to be "mine" in every present moment.

Thus, we can see that most scientific research on the self is too broad. If the self is the sense of being the subject of experience, it should not be conflated with a wider range of experiences. "I" refers to the feeling that our faculties have been *appropriated*, that a center of will and cognition interior to the body, somewhere behind the face, is doing the seeing, hearing, and thinking. And yet, in seeking to understand the self, many scientists study things such as spatial cognition, voluntary action, feelings of body ownership, and episodic memory. While these phenomena greatly influence our experience in each moment, they are not integral to the feeling that we call "I."

Consider the sense of body ownership. It must be produced, at least in part, by the integration of different streams of sensory information: We feel the position of our limbs in space; we see them at the appropriate locations in our visual field; and our experience of touching objects generally coincides with the sight of them coming into contact with our skin. An analogous synchrony occurs whenever we execute a volitional movement. No doubt our sense of body ownership is essential for our survival and for relating to others. Any loss or distortion of this sense can be profoundly disorienting. But disorienting to whom? When I am lying on the operating table, feeling the first effects of intravenous sedation, and find that I can no longer sense the position of my limbs in space, or even the existence of my body, who is it that has been deprived of these inputs? It is *I*—the (almost) ever-present subject of my experience. It should be obvious that no faculty of which I might be deprived, while I remain the subject experiencing the results of such deprivation, can be integral to the self—though it may be integral to my personhood in a wider sense.

Several findings in the neuroscientific literature drive a wedge between body ownership and the feeling of being a self. For instance, a person can lose the sense of owning a limb, a condition known as *somatoparaphrenia*. Conversely, a person's body image can encompass the limbs of others or even inanimate objects. Consider the famous "rubber hand illusion":

> Each of ten subjects was seated with their left arm resting upon a small table. A standing screen was positioned beside the arm to hide it from the subject's view and a life-sized rubber model of a left hand and arm was placed on the table directly in front of the subject. The subject sat with eyes fixed on the artificial hand while we used two small paintbrushes to stroke the rubber hand and the subject's hidden hand, synchronising the timing of the brushing as closely as possible. . . . Subjects experienced an illusion in which they seemed to feel the touch not of the hidden brush but that of the viewed brush, as if the rubber hand had sensed the touch.[8]

Amazingly, through the use of head-mounted video displays, this illusion can be extended to the entire body, yielding an experience of "body swapping."[9] It has long been known that vision trumps *proprioception* (the awareness of the position of one's body) when it comes to locating parts of one's body in space, but the "body swapping illusion" suggests that visual perception may fully determine the coordinates of the self.

The point, however, is that this effect—dissociation from one's own body and a false sense of inhabiting the parts (or whole body) of another person—seems to leave the "self" very much intact. Experiments on proprioception tell us nothing about the feeling that we call "I." And the same can be said about almost every other aspect of personhood with which philosophers, psychologists, and neuroscientists regularly bundle the self. The feeling of agency— the sense that one is the author of one's voluntary actions—may be as integral to our experience of the world as body ownership, but it, too, fails to capture what we mean by "self." A person could, for instance, distinguish his bodily movements from those of another person without feeling a sense of self at all, for to do so merely requires that he distinguish one body (as an object) from another. Likewise, he could fail to make such a distinction (in that he might misattribute his actions to another person or ascribe the actions of another to himself) while feeling the embrace of self-hood all the while.

Ascriptions of agency do not define the contours of the self in the way that many people seem to believe. While schizophrenics suffering from thought insertion, delusions of control, and auditory hallucinations[10] may be assailed by unusual mental phenomena, nothing suggests that their sense of *being a self* has been altered or lost. A person can fail to distinguish between self-generated and world-generated content, and thereby mistake her own internal imagery for sense data. There is a difference, to be sure, between finding a rat in one's bed and hallucinating (or merely dreaming about) such an encounter. But the feeling of being a self remains constant.

Self-Recognition

Imagine that you awake from a heavy sleep to find yourself imprisoned in an unfamiliar, windowless room. Where are you? You haven't the faintest idea. A mirror has been provided for your edification, however, and you gaze into it. What do you see? A red dot has been painted on your forehead, but for some reason you fail to notice it. In fact, you soon lose interest in your reflection altogether and begin searching your room for food. You are, after all, a gorilla, and quite unconcerned about your appearance.

In reviewing the literature on the self, one finds that much has been made of the fact that some creatures will attend to their reflections in a mirror with all the vanity of an eighteenth-century lady-in-waiting, while others respond as they would to a fellow member of their species.[11] The "mirror test" has been a staple of primate and child development research for many decades now, and it has made this simplest of all laboratory devices seem like a virtual dowsing rod for the self—because only those creatures who comport themselves with the requisite narcissism in front of the glass are believed to possess "self-knowledge" or even (and here we are treated to an especially depressing misuse of the term) "consciousness." While mirror self-recognition and use of the personal pronoun seem to emerge at more or less the same time in human development (fifteen to twenty-four months), there are many reasons to believe that self-recognition and selfhood are distinct states of mind—and, therefore, that they differ at the level of brain as well.[12]

Self-recognition depends on context. There are neurological patients who cannot recognize themselves in a mirror (a condition called the "mirror-sign delusion") but can pick themselves out in photographs,[13] and these subjects show no evidence of having lost anything like a *self*, or knowledge thereof. So what is the relationship between self-recognition and the feeling we call "I"? The fact that the word *self* is generally used while making reference to these phenomena does not suggest that any deep relationship exists between them. It seems quite possible, for instance, that a person who cannot recognize his own face under any circumstances could have a fully intact sense of self, just as your sense of self would remain unaltered by the sight of a complete stranger. There is simply nothing about the experience of *not recognizing a face*, even if it happens to be one's own, that suggests a divestiture of self or anything like it.

Theory of Mind

One of the most important things we do with our minds is attribute mental states to other people, a faculty that has been variously described as "theory of mind," "mentalizing," "mindsight," "mind reading," and the "intentional stance."[14] The ability to recognize and interpret the mental activity of others is essential for normal cognitive and social development, and deficits in this area contribute to a variety of mental disorders, including autism. But what is the relationship between an awareness of others and awareness of oneself? Many scientists and philosophers have suggested that the two must be deeply connected.[15] If so, it seems natural that research on theory of mind (TOM) would shed some light on the structure of the self. Unfortunately, however, the model of TOM that researchers generally work with cannot do this. Consider the following text, intended to evoke TOM processing in experimental subjects:

A burglar who has just robbed a shop is making his getaway. As he is running home, a policeman on his beat sees him drop his glove. He doesn't know the man is a burglar, he just wants to tell him he dropped his glove. But when the policeman shouts out to the burglar, "Hey, you! Stop!" the burglar turns round, sees the policeman and gives himself up. He puts his hands up and admits that he did the break-in at the local shop.

Question: Why did the burglar do that?[16]

The answer is obvious, unless one happens to be a young child or a person suffering from autism. If one can't take the point of view of the burglar in this story, it is impossible to know why he behaved as he did. Experimental stimuli of this kind are central to research on TOM, but they have very little to do with our most basic attribution of mindedness to others. Although we use our powers of inference to attribute complex mental states to other people, and the phrase "theory of mind" captures this, it seems that we make a much more basic attribution first, and perhaps independently: We recognize that other people are (or can be) *aware of us*. Explaining the burglar's behavior requires a higher level of cognition than is necessary to merely grasp that one is in the presence of a sentient *other*. And the feeling that another person can see or hear me is quite distinct from my having any understanding of his beliefs or desires. This more primitive judgment would seem to be TOM at its most fundamental. It might also have a deep connection to our sense of self.

The French philosopher Jean-Paul Sartre believed that our encounters with other people constitute the primal circumstance of self-formation.[17] On his account, each of us is perpetually in the position of a voyeur who, while gazing upon the object of his lust, suddenly hears the sound of someone stepping up directly behind him. Again and again, we are thrust out of the safety and seclusion of pure subjectivity by the knowledge that we have become objects in the world for others.

I believe that Sartre was onto something. The primitive impression that another creature is aware of us seems to be the point at which TOM is relevant to the sense of self. If you doubt this, I recommend the following exercise: Go to a public place, select a person at random, and stare at his face until he returns your gaze. To make this more than a pointless provocation, observe the change that occurs in you the moment eye contact is established. What is this feeling that obliges you to immediately look away or to begin speaking? The self-ramifying quality of this form of TOM seems indisputable, for without the attribution of awareness to others, you have no sense of being looked *at* in the first place. There is a difference to be felt here—being looked at just *feels* different from not being looked at—and the difference can be described, or so I maintain, as a magnification of the feeling that we call "I." It seems undeniable that self-consciousness and this more fundamental form of TOM are closely related.[18] The neurologist V. S. Ramachandran seems to have been thinking along these lines when he wrote, "It may not be coincidental that [you] use phrases like 'self conscious' when you really mean that you are conscious of others being conscious of you."[19]

To better appreciate the distinction between fundamental TOM and the TOM that is current in the scientific literature, consider what happens when we watch a film. The experience of sitting in a darkened theater and seeing people interact with one another on the screen is a social encounter of sorts—but it is one in which we, as participants, have been perfectly effaced. This very likely explains why most of us find movies and television so compelling. The moment we turn our eyes to the screen, we are in a social situation that our hominid genes could not have foreseen: We can view the actions of others, along with the minutiae of their facial expressions—even to the point of making eye contact with them—without the slightest risk of being observed ourselves. Movies and television magically transform the primordial context of face-to-face encounters, in which human beings have always been subjected to harrowing social lessons, allowing us, for the first time, to devote ourselves wholly to the act of observing other people. This is voyeurism of a transcendental kind. Whatever else might be said about the experience of watching a film, it fully dissociates fundamental TOM from standard TOM, for there is no doubt that we attribute mental states to the actors on the screen. We make all the judgments that the standard concept of TOM requires, but this does little to establish our sense of self. Indeed, it is difficult to find a situation in which we feel *less* self-conscious than when sitting in a darkened theater watching a film, and yet, we are contemplating the beliefs, intentions, and desires of other people the entire time.

Ramachandran and others have noted that the discovery of "mirror neurons" offers some support for the idea that the senses of self and other may emerge from the same circuitry in the brain. Some people believe that mirror neurons are also central to our ability to empathize with others and may even account for the emergence of gestural communication and spoken language. What we do know is that certain neurons increase their firing rate when we perform object-oriented actions with our hands (grasping, manipulating) and communicative or ingestive actions with our mouths. These neurons also fire, albeit less rapidly, whenever we witness the same actions performed by other people. Research on monkeys suggests that these neurons encode the *intentions* behind an observed action (such as picking up an apple for the purpose of eating it versus merely moving it) rather than the physical movements themselves. In these experiments, a monkey's brain seems to represent the purposeful behavior of others as if it were engaging in this behavior itself. Similar results have been obtained in neuroimaging experiments done on humans.[20]

Some scientists believe that mirror neurons provide a physiological basis for the development of imitation and social bonding early in life and for the understanding of other minds thereafter.[21] And it is certainly suggestive that children with autism appear to have diminished mirror neuron activity in proportion to the severity of their symptoms.[22] As is now widely known, people suffering from autism tend to lack insight into the mental lives of others. Conversely, a longitudinal study of compassion meditation, which produced a significant increase in subjects' empathy over the course of eight weeks, found increased activity in one of the regions believed to contain mirror neurons.[23]

It may be that an awareness of other minds is a necessary condition for an awareness of one's own. Of course, this does not suggest that the feeling we call "I" will disappear when we are alone. If our knowledge of self and other is truly indivisible, our awareness of others must be internalized early in life. In psychological terms, this certainly seems a plausible way of describing the structure of our subjectivity. All parents have seen their children put their growing powers of speech to use by maintaining running monologues with themselves. These monologues continue throughout life as though they were, in fact, *dialogues*. The resulting conversation seems both strange and unnecessary. Why should we live in *relationship to* ourselves rather than merely *as* ourselves? Why should an "I" and a "me" be keeping each other company?

Imagine that you have lost your sunglasses. You search the house up and down, and finally you spot them, lying on a table where you had left them the day before. You promptly think, "There they are!" as you make your way across the room to retrieve them. But to whom are you thinking this thought? You may even have uttered the phrase out loud: "There they are!" But who needed to be informed in this way? *You* have already seen them. Is there someone else in your search party?

Imagine that you are in a public place and happen to see a stranger locate his own lost sunglasses. He exclaims, as you might, "There they are!" and snatches them from the tabletop. A twinge of embarrassment often passes through all parties in such moments, but when the utterance is confined to a short phrase and occasioned by such an innocuous event, the speaker has done nothing out of the ordinary and bystanders are not yet gripped by fear. Imagine, however, if this person continued to address himself out loud: "Where did you think they were, you idiot? You've been wandering around this building for ten minutes. Now I'm going to be late for my lunch with Julie, and she's always on time!" The man need not speak another word to secure our eternal mistrust of his faculties. And yet the condition of this person is no different from our own—these are precisely the thoughts we might think in the privacy of our minds.

We have seen that the sense of self is logically and empirically distinct from many other features of the mind with which it is often conflated. In order to understand it at the level of the brain, therefore, we would need to study people who no longer experienced it. As we will see, certain practices of meditation are very well suited to research of this kind.

PENETRATING THE ILLUSION

As a matter of neurology, the sense of having a persistent and unified self must be an illusion, because it is built upon processes that, by their very nature as processes, are transitory and multifarious. There is no region of the brain that can be the seat of a soul. Everything that makes us human—our emotional lives, capacity for language, the impulses that give rise to complex behavior, and our ability to restrain other impulses that we consider uncivilized—is spread across the entirety of the cortex and many subcortical brain regions as well. The whole brain is involved in making us what we are. So we need not await any data from the lab to say that the self cannot be what it seems.

The sense that we are unified subjects is a fiction, produced by a multitude of separate processes and structures of which we are not aware and over which we exert no conscious control. What is more, many of these processes can be independently disturbed, producing deficits that would seem impossible if they were not so easily verified. Some people, for instance, are able to see perfectly but are unable to detect motion. Others are able to see objects and their motion but are unable to locate them in space. How the mind depends upon the brain, and the manner in which its powers can be disrupted, defies common sense. Here, as elsewhere in science, how things seem is often a poor guide to how they are.

The claim that we can experience consciousness without a conventional sense of self—that there is no rider on the horse—seems to be on firm ground neurologically. Whatever causes the brain to produce the false notion that there is a thinker living somewhere inside the head, it makes sense that it could stop doing this. And once it does, our inner lives become more faithful to the facts.

How can we know that the conventional sense of self is an illusion? When we look closely, it vanishes. This is compelling in the same way that the disappearance of any illusion is: You thought something was there, but upon closer inspection, you see that it isn't. What doesn't survive scrutiny cannot be real.

The classic example from the Indian tradition is of a coiled rope mistaken for a snake: Imagine that you spot a snake in the corner of a room and feel an immediate cascade of fear. But then you notice that it isn't moving. You look more closely and see that it doesn't appear to have a head—and suddenly you spot coiled strands of fiber that you mistook for a pattern of scales. You move closer and can see that it is a rope. A skeptic might ask, "How do you know that the rope is real and the snake an illusion?" This question may seem reasonable, but only to a person who hasn't had this experience of looking closely at the snake only to have it disappear. Given that the snake *always* collapses into being a rope, and not the other way around, there is simply no empirical basis upon which to form such a doubt.

Perhaps you can see the same effect in the above illusion. It certainly *looks* like there is a white square in the center of the figure, but when we study the image, it becomes clear that there are only four partial circles. The square has been imposed by our visual system, whose edge detectors have been fooled. Can we *know* that the black shapes are more real than the white square? Yes, because the square doesn't survive our efforts to locate it—its edges literally disappear. A little investigation and we see that its form has been merely implied. In fact, it is possible to look closely enough at the figure to banish the illusion altogether. But what could we say to a skeptic who insisted that the white square is just as real as the three-quarter circles? All we could do is urge him to look more closely. This is not a matter of debating third-person facts; it is a matter of looking more closely at experience itself.

In the next chapter, we will see that the illusion of the self can be investigated—and dispelled—in just this way.

Chapter 4

Meditation

Psychologists and neuroscientists now acknowledge that the human mind tends to wander, engaging in what has been called "stimulus-independent thought." The primary method of studying mental phenomena of this kind outside the lab is a technique called "experience sampling." Using a mobile phone or some other device, subjects are simply prompted to describe what they are doing and how they feel at random intervals throughout the day. One study found that when asked whether their mind was wandering—that is, whether they were thinking about something unrelated to their current experience—subjects reported being lost in thought 46.9 percent of the time.[1] Anyone who has trained in meditation will know that the true figure is surely higher—especially if we were to count all the thinking that, while perhaps superficially related to the task at hand, nevertheless constitutes an unnecessary distraction from it. As unreliable as such self-reports must be, this study found that people are consistently less happy when their minds are wandering, even when the contents of their thoughts are pleasant. The authors concluded that "a human mind is a wandering mind, and a wandering mind is an unhappy mind." Anyone who has spent time on silent retreat will agree.

The wandering mind has been correlated with activity in the brain's midline regions, especially the medial prefrontal cortex and the medial parietal cortex. These areas are often called the "default-mode" or "resting state" network because they are most active when we are just biding our time, waiting for something to happen. Activity in the default-mode network (DMN) decreases when subjects concentrate on tasks of the sort employed in most neuroimaging experiments.[2]

The DMN has also been linked with our capacity for "self-representation."[3] For instance, if a person believes that she is tall, the term *tall* should yield a greater signal in these midline regions than the term *short*. Similarly, the DMN is more engaged when we make such judgments of relevance about ourselves, as opposed to making them about other people. It also tends to be more active when we evaluate a scene from a first-person (rather than third-person) point of view.[4]

DEFAULT MODE NETWORK
(Midline Components)

Medial Prefrontal
Cortex

Medial Parietal
Cortex

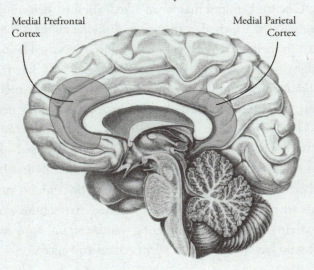

Generally speaking, to pay attention outwardly reduces activity in the brain's midline, while thinking about oneself increases it. These results appear mutually reinforcing and might explain the common experience we have "losing ourselves in our work."[5] Mindfulness and loving-kindness meditation (Pali: *metta*) also decrease activity in the DMN—and the effect is most pronounced among experienced meditators (both while meditating and at rest).[6] While it is too early to draw strong conclusions from these findings, they hint at a physical connection between the experience of being lost in thought and the sense of self (as well as a mechanism by which meditation might reduce both).

Long-term meditation practice is also associated with a variety of structural changes in the brain. Meditators tend to have larger corpora collosa and hippocampi (in both hemispheres). The practice is also linked to increased gray matter thickness and cortical folding. Some of these differences are especially prominent in older practitioners, which suggests that meditation could protect against age-related thinning of the cortex.[7] The cognitive, emotional, and behavioral significance of these anatomical findings have not yet been worked out, but it is not hard to see how they might explain the kinds of experiences and psychological changes that meditators report.

Expert meditators (with greater than ten thousand hours of practice) respond differently to pain than novices do. They judge the intensity of an unpleasant stimulus the same but find it to be less unpleasant. They also show reduced activity in regions associated with anxiety while anticipating the onset of pain, as well as faster habituation to the stimulus once it arrives.[8] Other research has found that mindfulness reduces both the unpleasantness and intensity of noxious stimuli.[9]

It has long been known that stress, especially early in life, alters brain structure. For instance, studies both in animals and in humans have shown that early stress increases the size of the amygdalae. One study found that an eight-week program of mindfulness meditation reduced the volume of the right basolateral amygdala, and these changes were correlated with a subjective decrease in stress.[10] Another found that a full day of mindfulness practice (among trained meditators) reduced the expression of several genes that produce inflammation throughout the body, and this correlated with an improved response to social stress (diabolically, subjects were asked to give a brief speech and then perform mental calculations while being videotaped in front of an audience).[11] A mere five minutes of practice a day (for five weeks) increased left-sided baseline activity in the frontal cortex—a pattern that, as we saw in the discussion of the split brain, has been associated with positive emotions.[12]

A review of the psychological literature suggests that mindfulness in particular fosters many components of physical and mental health: It improves immune function, blood pressure, and cortisol levels; it reduces anxiety, depression, neuroticism, and emotional reactivity. It also leads to greater behavioral regulation and has shown promise in the treatment of addiction and eating disorders. Unsurprisingly, the practice is associated with increased subjective well-being.[13] Training in compassion meditation increases empathy, as measured by the ability to accurately judge the emotions of others,[14] as well as positive affect in the presence of suffering.[15] The practice of mindfulness has been shown to have similar pro-social effects.[16]

Scientific research on the various types of meditation is just beginning, but there are now hundreds of studies suggesting that these practices are good for us. Again, from a first-person point of view, none of this is surprising. After all, there is an enormous difference between being hostage to one's thoughts and being freely and nonjudgmentally aware of life in the present. To make this shift is to interrupt the processes of rumination and reactivity that often keep us so desperately at odds with ourselves and with other people. No doubt many distinct mechanisms are involved—the regulation of attention and behavior, increased body awareness, inhibition of negative emotions, conceptual reframing of experience, changes in the view of "self," and so forth—and each of these processes will have its own neurophysiological causes. In the broadest sense, however, meditation is simply the ability to stop suffering in many of the usual ways, if only for a few moments at a time. How could that not be a skill worth cultivating?

GRADUAL VERSUS SUDDEN REALIZATION

We wouldn't attempt to meditate, or engage in any other contemplative practice, if we didn't feel that something about our experience needed to be improved. But here lies one of the central paradoxes of spiritual life, because this very feeling of dissatisfaction causes us to overlook the intrinsic freedom of consciousness in the present. As we have seen, there are good reasons to believe that adopting a practice like meditation can lead to positive changes in one's life. But the deepest goal of spirituality is freedom from the illusion of the self—and to *seek* such freedom, as though it were a future state to be attained through effort, is to reinforce the chains of one's apparent bondage in each moment.

Traditionally, there have been two solutions to this paradox. One is to simply ignore it and adopt various techniques of meditation in the hope that a breakthrough will occur. Some people appear to succeed at this, but many fail. It is true that good things often happen in the meantime: We can become happier and more concentrated. But we can also despair of the whole project. The words of the sages may begin to sound like empty promises, and we are left hoping for transcendent experiences that never arrive or prove merely temporary.

The ultimate wisdom of enlightenment, whatever it is, cannot be a matter of having fleeting experiences. The goal of meditation is to uncover a form of well-being that is inherent to the nature of our minds. It must, therefore, be available in the context of ordinary sights, sounds, sensations, and even thoughts. Peak experiences are fine, but real freedom must be coincident with normal waking life.

The other traditional response to the paradox of spiritual seeking is to fully acknowledge it and concede that all efforts are doomed, because the urge to attain self-transcendence or any other mystical experience is a symptom of the very disease we want to cure. There is nothing to do but give up the search.

These paths may appear antithetical—and they are often presented as such. The path of gradual ascent is typical of Theravada Buddhism and most other approaches to meditation in the Indian tradition. And gradualism is the natural starting point for any search, spiritual or otherwise. Such goal-oriented modes of practice have the virtue of being easily taught, because a person can begin them without having had any fundamental insight into the nature of consciousness or the illusoriness of the self. He need only adopt new patterns of attention, thought, and behavior, and the path will unfold before him.

By contrast, the path of sudden realization can appear impossibly steep. It is often described as "nondualistic" because it refuses to validate the point of view from which one would meditate or practice any other spiritual discipline. Consciousness is already free of anything that remotely resembles a self—and there is nothing that *you* can do, as an illusory ego, to realize this. Such a perspective can be found in the Indian tradition of Advaita Vedanta and in a few schools of Buddhism.

Those who begin to practice in the spirit of gradualism often assume that the goal of self-transcendence is far away, and they may spend years overlooking the very freedom that they yearn to realize. The liability of this approach became clear to me when I studied under the Burmese meditation master Sayadaw U Pandita. I sat through several retreats with U Pandita, each a month or two in length. These retreats were based on the monastic discipline of Theravadan Buddhism: We did not eat after noon and were encouraged to sleep no more than four hours each night. Outwardly, the goal was to engage in eighteen hours of formal meditation each day. Inwardly, it was to follow the stages of insight as laid out in Buddhaghosa's fifth-century treatise, the *Visuddhimagga*, and elaborated in the writings of U Pandita's own legendary teacher, Mahasi Sayadaw.[17]

The logic of this practice is explicitly goal-oriented: According to this view, one practices mindfulness not because the intrinsic freedom of consciousness can be fully realized in the present but because being mindful is a means of attaining an experience often described as "cessation," which is thought to decisively uproot the illusion of the self (along with other mental afflictions, depending on one's stage of practice). Cessation is believed to be a direct insight into an unconditioned reality (Pali: *Nibbāna*; Sanskrit: *Nirvana*) that lies behind all manifest phenomena.

This conception of the path to enlightenment is open to several criticisms. The first is that it is misleading with respect to what can be realized in the present moment in a state of ordinary awareness. Thus, it encourages confusion at the outset regarding the nature of the problem one is trying to solve. It is true, however, that striving toward the distant goal of enlightenment (as well as the nearer goal of cessation) can lead one to practice with an intensity that might otherwise be difficult to achieve. I never made more effort than I did when practicing under U Pandita. But most of this effort arose from the very illusion of bondage to the self that I was seeking to overcome. The model of this practice is that one must climb the mountain so that freedom can be found at the top. But the self is *already* an illusion, and that truth can be glimpsed directly, at the mountain's base or anywhere else along the path. One can then return to this insight, again and again, as one's sole method of meditation—thereby arriving at the goal in each moment of actual practice.

This isn't merely a matter of choosing to think differently about the significance of mindfulness. It is a difference in what one is able to be mindful of. Dualistic mindfulness—paying attention to the breath, for instance—generally proceeds on the basis of an illusion: One feels that one is a subject, a locus of consciousness inside the head, that can strategically pay attention to the breath or some other object of awareness because of all the good it will do. This is gradualism in action. And yet, from a nondualistic point of view, one could just as well be mindful of selflessness directly. To do this, however, one must recognize that this is how consciousness is—and such an insight can be difficult to achieve. However, it does not require the meditative attainment of cessation. Another problem with the goal of cessation is that most traditions of Buddhism do not share it, and yet they produce long lineages of contemplative masters, many of whom have spent decades doing nothing but meditating on the nature of consciousness. If freedom is possible, there must be some mode of ordinary consciousness in which it can be expressed. Why not realize this frame of mind directly?

Nevertheless, I spent several years deeply preoccupied with reaching the goal of cessation, and at least one year of that time was spent on silent retreat. Although I had many interesting experiences, none seemed to fit the specific requirements of this path. There were periods during which all thought subsided, and any sense of having a body disappeared. What remained was a blissful expanse of conscious peace that had no reference point in any of the usual sensory channels. Many scientists and philosophers believe that consciousness is always tied to one of the five senses—and that the idea of a "pure consciousness" apart from seeing, hearing, smelling, tasting, and touching is a category error and a spiritual fantasy. I am confident that they are mistaken.

But cessation never arrived. Given my gradualist views at that point, this became very frustrating. Most of my time on retreat was extremely pleasant, but it seemed to me that I had merely been given the tools with which to contemplate the evidence of my nonenlightenment. My practice had become a vigil—a method of waiting, however patiently, for a future reward.

The pendulum swung when I met an Indian teacher named H. W. L. Poonja (1910–97), called "Poonja-ji" or "Papaji" by his students. Poonja-ji was a disciple of Ramana Maharshi (1879–1950), arguably the most widely revered Indian sage of the twentieth century. Ramana's own awakening had been quite unusual, because he had no apparent spiritual interests or contact with a teacher. As a boy of sixteen, living in a middle-class family of South Indian Brahmins, he spontaneously became a spiritual adept.

While sitting alone in his uncle's study, Ramana suddenly became paralyzed by a fear of death. He lay down on the floor, convinced that he would soon die, but rather than remaining terrified, he decided to locate the self that was about to disappear. He focused on the feeling of "I"—a process he later called "self-inquiry"—and found it to be absent from the field of consciousness. Ramana the person didn't die that day, but he claimed that the feeling of being a separate self never darkened his consciousness again.

After fruitlessly attempting to behave like the ordinary boy he had once been, Ramana left home and traveled to Tiruvannamalai, an ancient pilgrimage site for followers of Shiva. He spent the rest of his life there, in proximity to the mountain Arunachala, with which he claimed to have a mystical connection.

In the early years after his awakening, Ramana seemed to lose his ability to speak, and he was said to grow so absorbed in his experience of transfigured consciousness that he remained motionless for days at a time. His body grew weak, developed sores, and had to be tended by the few locals who had taken an interest in him. After a decade of silence, around 1906, Ramana began to conduct dialogues about the nature of consciousness. Until the end of his life, a steady stream of students came to study with him. These are the sorts of things he was apt to say:

> The mind is a bundle of thoughts. The thoughts arise because there is the thinker. The thinker is the ego. The ego, if sought, will automatically vanish.[18]

> Reality is simply the loss of the ego. Destroy the ego by seeking its identity. Because ego is no entity it will automatically vanish and reality will shine forth by itself. This is the direct method, whereas all other methods are done, only retaining the ego. . . . No *sadhanas* [spiritual practices] are necessary for engaging in this quest.

There is no greater mystery than this—that being the reality we seek to gain reality. We think that there is something hiding our reality and that it must be destroyed before the reality is gained. It is ridiculous. A day will dawn when you will yourself laugh at your past efforts. That which will be on the day you laugh is also here and now.[19]

Any attempt to make sense of such teachings in third-person, scientific terms quickly produces monstrosities. From the point of view of psychological science, for instance, the mind is not just "a bundle of thoughts." And in what sense can reality be "simply the loss of the ego"? Does this reality include quasars and hantavirus? But these are the kinds of quibbles that will cause one to miss Ramana's point.

While the philosophy of Advaita, and Ramana's own words, may tend to support a metaphysical reading of teachings of this kind, their validity is not metaphysical. Rather, it is experiential. The whole of Advaita reduces to a series of very simple and testable assertions: Consciousness is the prior condition of every experience; the self or ego is an illusory appearance within it; look closely for what you are calling "I," and the feeling of being a separate self will disappear; what remains, as a matter of experience, is a field of consciousness—free, undivided, and intrinsically uncontaminated by its ever-changing contents.

These are the simple truths that Poonja-ji taught. In fact, he was even more uncompromising than his guru in his nonduality. Whereas Ramana would often concede the utility of certain dualistic practices, Poonja-ji never gave an inch. The effect was intoxicating, especially to those of us who had spent years practicing meditation. Poonja-ji was also given to spontaneous bouts of weeping and laughter—both, apparently, from sheer joy. The man did not hide his light under a bushel. When I first met him, he had not yet been discovered by the throngs of Western devotees who would soon turn his tiny house in Lucknow into a spiritual circus. Like his teacher Ramana, Poonja-ji claimed to be perfectly free from the illusion of the self—and by all appearances, he was. And like Ramana—and every other Indian guru—Poonja-ji would occasionally say something deeply unscientific. On the whole, however, his teaching was remarkably free of Hindu religiosity or unwarranted assertions about the nature of the cosmos. He appeared to simply speak from experience about the nature of experience itself.

Poonja-ji's influence on me was profound, especially because it came as a corrective to all the strenuous and unsatisfying efforts I had been making in meditation up to that point. But the dangers inherent in his approach soon became obvious. The all-or-nothing quality of Poonja-ji's teaching obliged him to acknowledge the full enlightenment of any person who was grandiose or manic enough to claim it. Thus, I repeatedly witnessed fellow students declare their complete and undying freedom, all the while appearing quite ordinary—or worse. In certain cases, these people had clearly had some sort of breakthrough, but Poonja-ji's insistence upon the finality of every legitimate insight led many of them to delude themselves about their spiritual attainments. Some left India and became gurus. From what I could tell, Poonja-ji gave everyone his blessing to spread his teachings in this way. He once suggested that I do it, and yet it was clear to me that I was not qualified to be anyone's guru. Nearly twenty years have passed, and I'm still not. Of course, from Poonja-ji's point of view, this is an illusion. And yet there simply is a difference between a person like myself, who is generally distracted by thought, and one who isn't and cannot be. I don't know where to place Poonja-ji on this continuum of wisdom, but he appeared to be a lot farther along than his students. Whether Poonja-ji was capable of seeing the difference between himself and other people, I do not know. But his insistence that no difference existed began to seem either dogmatic or delusional.

On one occasion, events conspired to perfectly illuminate the flaw in Poonja-ji's teaching. A small group of experienced practitioners (among us several teachers of meditation) had organized a trip to India and Nepal to spend ten days with Poonja-ji in Lucknow, followed by ten days in Kathmandu, to receive teachings on the Tibetan Buddhist practice of Dzogchen. As it happened, during our time in Lucknow, a woman from Switzerland became "enlightened" in Poonja-ji's presence. For the better part of a week, she was celebrated as something akin to the next Buddha. Poonja-ji repeatedly put her forward as evidence of how fully the truth could be realized without making any effort at all in meditation, and we had the pleasure of seeing this woman sit beside Poonja-ji on a raised platform expounding upon how blissful it now was in her corner of the universe. She was, in fact, radiantly happy, and it was by no means clear that Poonja-ji had made a mistake in recognizing her. She would say things like "There is nothing but consciousness, and there is no difference between it and reality itself." Coming from such a nice, guileless person, there was little reason to doubt the profundity of her experience.

When it came time for our group to leave India for Nepal, this woman asked if she could join us. Because she was such good company, we encouraged her to come along. A few of us were also curious to see how her realization would appear in another context. And so it came to pass that a woman whose enlightenment had just been confirmed by one of the greatest living exponents of

Advaita Vedanta was in the room when we received our first teach-ings from Tulku Urgyen Rinpoche, who was generally thought to be one of the greatest living Dzogchen masters.

Of all the Buddhist teachings, those of Dzogchen most closely resemble the teachings of Advaita. The two traditions seek to pro-voke the same insight into the nonduality of consciousness, but, generally speaking, only Dzogchen makes it absolutely clear that one must *practice* this insight to the point of stability and that one can do so without succumbing to the dualistic striving that haunts most other paths.

At a certain point in our discussions with Tulku Urgyen, our Swiss prodigy declared her boundless freedom in terms similar to those she had used to such great effect with Poonja-ji. After a few highly amusing exchanges, during which we watched Tulku Ur-gyen struggle to understand what our translator was telling him, he gave a short laugh and looked the woman over with renewed interest.

"How long has it been since you were last lost in thought?" he asked.

"I haven't had any thoughts for over a week," the woman replied.

Tulku Urgyen smiled.

"A week?"

"Yes."

"No thoughts?"

"No, my mind is completely still. It's just pure con-sciousness."

"That's very interesting. Okay, so this is what is going to happen now: We are all going to wait for you to have your next thought. There's no hurry. We are all very patient people. We are just going to sit here and wait. Please tell us when you notice a thought arise in your mind."

It is difficult to convey what a brilliant and subtle intervention this was. It may have been the most inspired moment of teaching I have ever witnessed.

After a few moments, a look of doubt appeared on our friend's face.

"Okay . . . Wait a minute . . . Oh . . . That could have been a thought there . . . Okay . . ."

Over the next thirty seconds, we watched this woman's enlightenment completely unravel. It became clear that she had been merely *thinking* about how expansive her experience of consciousness had become—how it was perfectly free of thought, immaculate, just like space—without noticing that she was *thinking incessantly.* She had been telling herself the story of her enlightenment—and she had been getting away with it because she happened to be an extraordinarily happy person for whom everything was going very well for the time being.

This was the danger of nondual teachings of the sort that Poonja-ji was handing out to all comers. It was easy to delude oneself into thinking that one had achieved a permanent breakthrough, especially because he insisted that all breakthroughs must be permanent. What the Dzogchen teachings make clear, however, is that thinking about what is beyond thought is still thinking, and a glimpse of selflessness is generally only the beginning of a process that must reach fruition. Being able to stand perfectly free of the feeling of self is the *start* of one's spiritual journey, not its end.

DZOGCHEN: TAKING THE GOAL AS THE PATH

Tulku Urgyen Rinpoche lived in a hermitage on the southern slope of Shivapuri Mountain, overlooking the Kathmandu Valley. He spent more than twenty years of his life on formal retreat and was deservedly famous for the clarity with which he gave the "pointing-out instruction" of Dzogchen, a formal initiation in which a teacher seeks to impart the experience of self-transcendence directly to a student. I received this teaching from several Dzogchen masters, as well as similar instructions from teachers like Poonja-ji in other traditions, but I never met anyone who spoke about the nature of consciousness as precisely as Tulku Urgyen. In the last five years of his life, I made several trips to Nepal to study with him.

The practice of Dzogchen requires that one be able to experience the intrinsic selflessness of awareness in every moment (that is, when one is not otherwise distracted by thought)—which is to say that for a Dzogchen meditator, mindfulness must be synonymous with dispelling the illusion of the self. Rather than teach a technique of meditation—such as paying close attention to one's breathing—a Dzogchen master must precipitate an insight on the basis of which a student can thereafter practice a form of awareness (Tibetan: *rigpa*) that is unencumbered by subject/object dualism. Thus, it is often said that, in Dzogchen, one "takes the goal as the path," because the freedom from self that one might otherwise seek is the very thing that one practices. The goal of Dzogchen, if one can call it such, is to grow increasingly familiar with this way of being in the world.

In my experience, some Dzogchen masters are better teachers than others. I have been in the presence of several of the most revered Tibetan lamas of our time while they were ostensibly teaching Dzogchen, and most of them simply described this view of consciousness without giving clear instructions on how to glimpse it. The genius of Tulku Urgyen was that he could point out the nature of mind with the precision and matter-of-factness of teaching a person how to thread a needle and could get an ordinary meditator like me to recognize that consciousness is intrinsically free of self. There might be some initial struggle and uncertainty, depending on the student, but once the truth of nonduality had been glimpsed, it became obvious that it was always available— and there was never any doubt about how to see it again. I came to Tulku Urgyen yearning for the experience of self-transcendence, and in a few minutes he showed me that I had no self to transcend.

In my view, there is nothing supernatural, or even mysterious, about this transmission of wisdom from master to disciple. Tulku Urgyen's effect on me came purely from the clarity of his teaching. As it is with any challenging endeavor, the difference between being utterly misled by false information, being nudged in the general direction, and being precisely guided by an expert is difficult to overstate.

The direct perception of the optic blind spot again provides a useful analogy: Imagine that perceiving the blind spot will completely transform a person's life. Next, imagine that whole religions such as Judaism, Christianity, and Islam are predicated on the denial of the blind spot's existence—let us say that their central doctrines assert the perfect uniformity of the visual field. Perhaps other traditions acknowledge the blind spot but in purely poetical terms, without giving any clear indication of how to recognize it. A few lineages may actually teach techniques whereby one can see the blind spot for oneself, but only gradually, after months and years of effort, and even then one's glimpses of it will seem more a matter of luck than anything else. In a more esoteric tradition still, a "blind spot master" gives the "pointing-out instruction" but without much precision: Perhaps he tells you to close one eye, for reasons that are never made explicit, and then says that the spot you seek is right on the surface of your vision. No doubt some people will succeed in discovering the blind spot under these conditions, but the teacher could certainly be clearer than this. How much clearer? If Tulku Urgyen had been pointing out the blind spot, he would have produced a figure like the one below and given these instructions:

1. Hold this figure in front of you at arm's length.
2. Close your left eye and stare at the cross with your right.
3. Gradually move the page closer to your face while keeping your gaze fixed on the cross.
4. Notice when the dot on the right disappears.
5. Once you find your blind spot, continue to experiment with this figure by moving the page back and forth until any possibility of doubt about the existence of the blind spot has disappeared.

         ●

It is considered bad form in most spiritual circles, especially among Buddhists, to make claims about one's own realization. However, I think this taboo comes at a high price, because it allows people to remain confused about how to practice. So I will describe my experience plainly.

Before meeting Tulku Urgyen, I had spent at least a year practicing *vipassana* on silent retreats. The experience of self-transcendence was not entirely unknown to me. I could remember moments when the distance between the observer and the observed had seemed to vanish, but I viewed these experiences as being dependent on conditions of extreme mental concentration. Consequently, I thought they were unavailable in more ordinary moments, outside intensive retreat. But after a few minutes, Tulku Urgyen simply handed me the ability to cut through the illusion of the self directly, even in ordinary states of consciousness. This instruction was, without question, the most important thing I have ever been explicitly taught by another human being. It has given me a way to escape the usual tides of psychological suffering—fear, anger, shame—in an instant. At my level of practice, this freedom lasts only a few moments. But these moments can be repeated, and they can grow in duration. Punctuating ordinary experience in this way makes all the difference. In fact, when I pay attention, it is impossible for me to feel like a self at all: The implied center of cognition and emotion simply falls away, and it is obvious that consciousness is never truly confined by what it knows. That which is aware of sadness is not sad. That which is aware of fear is not fearful. The moment I am lost in thought, however, I'm as confused as anyone else.

Given this change in my perception of the world, I understand the attractions of traditional spirituality. I also recognize the needless confusion and harm that inevitably arise from the doctrines of faith-based religion. I did not have to believe anything irrational about the universe, or about my place within it, to learn the practice of Dzogchen. I didn't have to accept Tibetan Buddhist beliefs about karma and rebirth or imagine that Tulku Urgyen or the other meditation masters I met possessed magic powers. And whatever the traditional liabilities of the guru-devotee relationship, I know from direct experience that it is possible to meet a teacher who can deliver the goods.

Unfortunately, to begin the practice of Dzogchen, it is generally necessary to meet a qualified teacher. There is a large literature on the topic, of course, and much of what I have written throughout this book represents my own effort to "point out" the nature of awareness. But to have their confusion and doubts resolved, most people need to be in a dialogue with a teacher who can answer questions in real time. Tulku Urgyen is no longer alive, but I'm told that his sons Tsoknyi Rinpoche and Mingyur Rinpoche generally teach in his style, and many other Tibetan lamas teach Dzogchen as well. However, one can never be sure how much Buddhist religiosity one will be asked to imbibe along the way. My advice is that if you seek out these teachings, don't be satisfied until you are certain that you understand the practice. Dzogchen is not vague or paradoxical. It is not like Zen, wherein a person can spend years being uncertain whether he is meditating correctly. The practice of recognizing nondual awareness is called *trekchod*, which means "cutting through" in Tibetan, as in cutting a string cleanly so that both ends fall away. Once one has cut it, there is no doubt that it has been cut. I recommend that you demand the same clarity of your meditation practice.

Beyond Duality

Think of something pleasant in your personal life—visualize the moment when you accomplished something that you are proud of or had a good laugh with a friend. Take a minute to do this. Notice how the mere thought of the past evokes a feeling in the present. But does consciousness itself feel happy? Is it truly changed or colored by what it knows?

In the teachings of Dzogchen, it is often said that thoughts and emotions arise in consciousness the way that images appear on the surface of a mirror. This is only a metaphor, but it does capture an insight that one can have about the nature of the mind. Is a mirror improved by beautiful images? No. The same can be said for consciousness.

Now think of something unpleasant: Perhaps you recently embarrassed yourself or received some bad news. Maybe there is an upcoming event about which you feel acutely anxious. Notice whatever feelings arise in the wake of these thoughts. They are also appearances in consciousness. Do they have the power to change what consciousness is in itself?

There is real freedom to be found here, but you are unlikely to find it without looking carefully into the nature of consciousness, again and again. Notice how thoughts continue to arise. Even while reading this page your attention has surely strayed several times. Such wanderings of mind are the primary obstacle to meditation. Meditation doesn't entail the suppression of such thoughts, but it does require that we notice thoughts as they emerge and recognize them to be transitory appearances in consciousness. In subjective terms, you are consciousness itself—you are not the next, evanescent image or string of words that appears in your mind. Not seeing it arise, however, the next thought will seem to become what you are.

But how could you actually be a thought? Whatever their content, thoughts vanish almost the instant they appear. They are like sounds, or fleeting sensations in your body. How could this next thought define your subjectivity at all?

It may take years of observing the contents of consciousness—or it may take only moments—but it is quite possible to realize that consciousness itself is free, no matter what arises to be noticed. Meditation is the practice of finding this freedom directly, by breaking one's identification with thought and allowing the continuum of experience, pleasant and unpleasant, to simply be as it is. There are many traditional techniques for doing this. But it is important to realize that true meditation isn't an effort to produce a certain state of mind—like bliss, or unusual visual images, or love for all sentient beings. Such methods also exist, but they serve a more limited function. The deeper purpose of meditation is to recognize that which is common to all states of experience, both pleasant and unpleasant. The goal is to realize those qualities that are intrinsic to consciousness in every present moment, no matter what arises to be noticed.

When you are able to rest naturally, merely witnessing the totality of experience, and thoughts themselves are left to arise and vanish as they will, you can recognize that consciousness is intrinsically undivided. In the moment of such an insight, you will be completely relieved of the feeling that you call "I." You will still see this book, of course, but it will be an appearance in consciousness, inseparable from consciousness itself—and there will be no sense that you are behind your eyes, doing the reading.

Such a shift in view isn't a matter of thinking new thoughts. It is easy enough to *think* that this book is just an appearance in consciousness. It is another matter to recognize it as such, prior to the arising of thought.

The gesture that precipitates this insight for most people is an attempt to invert consciousness upon itself—to look for that which is looking—and to notice, in the *first* instant of looking for your self, what happens to the apparent divide between subject and object. Do you still feel that you are over there, behind your eyes, looking out at a world of objects?

It really is possible to look for the feeling you are calling "I" and to fail to find it in a way that is conclusive.

HAVING NO HEAD

Douglas Harding was a British architect who later in life became celebrated in New Age circles for having opened a novel doorway into the experience of selflessness. Raised among the Exclusive Plymouth Brethren, a highly repressive sect of fundamentalist Christians, Harding apparently expressed his doubts with a fervor sufficient to get himself excommunicated for apostasy. He later moved his family to India, where he spent years on a journey of self-discovery that culminated in an insight he described as the state of "having no head." I never met Harding, but after reading his books, I have little doubt that he was attempting to introduce his students to the same understanding that is the basis of Dzogchen practice.

Harding was led to his insight after seeing a self-portrait of the Austrian physicist and philosopher Ernst Mach, who had the clever idea of drawing himself as he appeared from a first-person point of view: "I lie upon my sofa. If I close my right eye, the picture represented in the accompanying cut is presented to my left eye. In a frame formed by the ridge of my eyebrow, by my nose, and by my moustache, appears a part of my body, so far as visible, with its environment."[20] Harding later wrote several books about his experience, including a very useful little volume titled *On Having No Head*. It is both amusing and instructive to note that his teachings were singled out for derision by the cognitive scientist Douglas Hofstadter (in collaboration with my friend Daniel Dennett), a man of wide learning and great intelligence who, it would appear, did not understand what Harding was talking about.

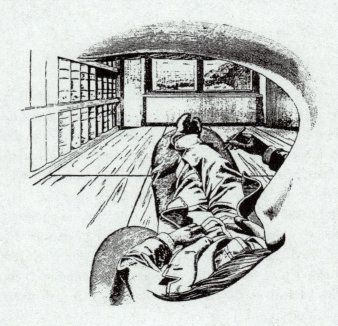

Here is a portion of Harding's text that Hofstadter criticized:

What actually happened was something absurdly simple and unspectacular: I stopped thinking. A peculiar quiet, an odd kind of alert limpness or numbness, came over me. Reason and imagination and all mental chatter died down. For once, words really failed me. Past and future dropped away. I forgot who and what I was, my name, manhood, animal-hood, all that could be called mine. It was as if I had been born that instant, brand new, mindless, innocent of all memories. There existed only the Now, that present moment and what was clearly given in it. To look was enough. And what I found was khaki trouser legs terminating downwards in a pair of brown shoes, khaki sleeves terminating sideways in a pair of pink hands, and a khaki shirtfront terminating upwards in absolutely nothing whatsoever! Certainly not in a head.

It took me no time at all to notice that this nothing, this hole where a head should have been, was no ordinary vacancy, no mere nothing. On the contrary, it was very much occupied. It was a vast emptiness vastly filled, a nothing that found room for everything: room for grass, trees, shadowy distant hills, and far above them snow-peaks like a row of angular clouds riding the blue sky. I had lost a head and gained a world. . . . Here it was, this superb scene, brightly shining in the clear air, alone and unsupported, mysteriously suspended in the void, and (and this was the real miracle, the wonder and delight) utterly free of "me," unstained by any observer. Its total presence was my total absence, body and soul. Lighter than air, clearer than glass, altogether released from myself. I was nowhere around. . . . There arose no questions, no reference beyond the experience itself, but only peace and a quiet joy, and the sensation of having dropped an intolerable burden. . . . I had been blind to the one thing that is always present, and without which I am blind

indeed to this marvelous substitute-for-a-head, this un-
bounded clarity, this luminous and absolutely pure void,
which nevertheless is—rather than contains—all things.
For, however carefully I attend, I fail to find here even so
much as a blank screen on which these mountains and
sun and sky are projected, or a clear mirror in which they
are reflected, or a transparent lens or aperture through
which they are viewed, still less a soul or a mind to which
they are presented, or viewer (however shadowy) who is
distinguishable from the view. Nothing whatever inter-
venes, not even that baffling and elusive obstacle called
"distance": the huge blue sky, the pink-edged whiteness
of the snows, the sparkling green of the grass—how can
these be remote, when there's nothing to be remote from?
The headless void refuses all definition and location: it is
not round, or small, or big, or even here as distinct from
there. [21]

Harding's assertion that he has no head must be read in the first-
person sense; the man was not claiming to have been literally
decapitated. From a first-person point of view, his emphasis on
headlessness is a stroke of genius that offers an unusually clear description of what it's like to glimpse the nonduality of consciousness.

Here are Hofstadter's "reflections" on Harding's account: "We
have here been presented with a charmingly childish and solipsistic view of the human condition. It is something that, at an intellectual level, offends and appalls us: can anyone sincerely entertain
such notions without embarrassment? Yet to some primitive level
in us it speaks clearly. That is the level at which we cannot accept
the notion of our own death."[22] Having expressed his pity for batty

old Harding, Hofstadter proceeds to explain away his insights as a solipsistic denial of mortality—a perpetuation of the childish illusion that "I am a necessary ingredient of the universe." However, Harding's point was that "I" is not even an ingredient, necessary or otherwise, of *his own mind*. What Hofstadter fails to realize is that Harding's account contains a precise, empirical instruction: Look for whatever it is you are calling "I" without being distracted by even the subtlest undercurrent of thought—and notice what happens the moment you turn consciousness upon itself.

This illustrates a very common phenomenon in scientific and secular circles: We have a contemplative like Harding who, to the eye of anyone familiar with the experience of self-transcendence, has described it in a manner approaching perfect clarity; we also have a scholar like Hofstadter, a celebrated contributor to our modern understanding of the mind, who dismisses him as a child.

Before rejecting Harding's account as merely silly, you should investigate this experience for yourself.

Look for Your Head

As you gaze at the world around you, take a moment to look for your head. This may seem like a bizarre instruction. You might think, "Of course, I can't see my head. What's so interesting about that?" Not so fast. Simply look at the world, or at other people, and attempt to turn your attention in the direction of where you know your head to be. For instance, if you are having a conversation with another person, see if you can let your attention travel in the direction of the other person's gaze. He is looking at your face—and *you* cannot see your face. The only face present, from your point of view, belongs to the other person. But looking for yourself in this way can precipitate a sudden change in perspective, of the sort Harding describes.

Some people find it easier to trigger this shift in a slightly different way: As you are looking out at the world, simply imagine that you have no head.

Whichever method you choose, don't struggle with this exercise. It is not a matter of going deep within or of producing some extraordinary experience. The view of headlessness is right on the surface of consciousness and can be glimpsed the moment you attempt to turn about. Pay attention to how the world appears in the *first* instant, not after a protracted effort. Either you will see it immediately or you won't see it at all. And the resulting glimpse of open awareness will last only a moment or two before thoughts intervene. Simply repeat this glimpse, again and again, in as relaxed a way as possible, as you go about your day.

Once again, selflessness is not a "deep" feature of consciousness. It is right on the surface. And yet people can meditate for years without recognizing it. After I was introduced to the practice of Dzogchen, I realized that much of my time spent meditating had been a way of actively overlooking the very insight I was seeking.

How can something be right on the surface of experience and yet be difficult to see? I have already drawn an analogy to the optic blind spot. But other analogies may give a clearer sense of the subtle shift in attention that is required to see what is right before one's eyes.

We've all had the experience of looking through a window and suddenly noticing our own reflection in the glass. At that moment we have a choice: to use the window as a window and see the world beyond, or to use it as a mirror. It is extraordinarily easy to shift back and forth between these two views but impossible to truly focus on both simultaneously. This shift offers a very good analogy both for what it is like to recognize the illusoriness of the self for the first time and for why it can take so long to do it.

Imagine that you want to show another person how a window can also function like a mirror. As it happens, your friend has never seen this effect and is quite skeptical of your claims. You direct her attention to the largest window in your house, and although the conditions are perfect for seeing her reflection, she immediately becomes captivated by the world outside. *What a beautiful view! Who are your neighbors? Is that a redwood or a Douglas fir?* You begin to speak about there being two views and about the fact that your friend's reflection stands before her even now, but she notices only that the neighbor's dog has slipped out the front door and is now dashing down the sidewalk. In every moment, it is clear to you that your friend is staring directly through the image of her face without seeing it.

Of course, you could easily direct her attention to the surface of the window by touching the glass with your hand. This would be akin to the "pointing-out instruction" of Dzogchen. However, here the analogy begins to break down. It is very difficult to imagine someone's not being able to see her reflection in a window even after years of looking—but that is what happens when a person begins most forms of spiritual practice. Most techniques of meditation are, in essence, elaborate ways for looking *through* the window in the hope that if one only sees the world in greater detail, an image of one's true face will eventually appear. Imagine a teaching like this: *If you just focus on the trees swaying outside the window without distraction, you will see your true face.* Undoubtedly, such

an instruction would be an obstacle to seeing what could otherwise be seen directly. Almost everything that has been said or written about spiritual practice, even most of the teachings one finds
in Buddhism, directs a person's gaze to the world beyond the glass,
thereby confusing matters from the very beginning.

But one must start somewhere. And the truth is that most
people are simply too distracted by their thoughts to have the
selflessness of consciousness pointed out directly. And even if
they are ready to glimpse it, they are unlikely to understand its
significance. Harding confessed that many of his students recognized the state of "headlessness" only to say, "So what?" It is, in
fact, very difficult to deal with this "So what?" That is why certain
traditions, like Dzogchen, consider teachings about the intrinsic
nonduality of consciousness to be secret, reserving them for students who have spent considerable time practicing other forms of
meditation. On one level, the requirement that a person have mas-
tered other preliminary practices is purely pragmatic—for unless
she has the requisite concentration and mindfulness to actually
follow the teacher's instructions, she is liable to be lost in thought
and understand nothing at all. But there is another purpose to
withholding these nondual teachings: Unless a person has spent
some time seeking self-transcendence dualistically, she is unlikely
to recognize that the brief glimpse of selflessness is actually the
answer to her search. Having then said, "So what?" in the face of
the highest teachings, there is nothing for her to do but persist in
her confusion.

THE PARADOX OF ACCEPTANCE

It would seem that very few good things in life come from our accepting the present moment as it is. To become educated, we must

be motivated to learn. To master a sport requires that we continually improve our performance and overcome our resistance to physical exertion. To be a better spouse or parent, we often must make a deliberate effort to change ourselves. Merely accepting that we are lazy, distracted, petty, easily provoked to anger, and inclined to waste our time in ways that we will later regret is not a path to happiness.

And yet it is true that meditation requires total acceptance of what is given in the present moment. If you are injured and in pain, the path to mental peace can be traversed in a single step: Simply accept the pain as it arises, while doing whatever you need to do to help your body heal. If you are anxious before giving a speech, become willing to feel the anxiety fully, so that it becomes a meaningless pattern of energy in your mind and body. Embracing the contents of consciousness in any moment is a very powerful way of training yourself to respond differently to adversity. However, it is important to distinguish between accepting unpleasant sensations and emotions as a strategy—while covertly hoping that they will go away—and *truly* accepting them as transitory appearances in consciousness. Only the latter gesture opens the door to wisdom and lasting change. The paradox is that we can become wiser and more compassionate and live more fulfilling lives by refusing to be who we have tended to be in the past. But we must also relax, accepting things as they are in the present, as we strive to change ourselves.

Chapter 5

Gurus, Death, Drugs, and Other Puzzles

O ne of the first obstacles encountered along any contemplative path is the basic uncertainty about the nature of spiritual authority. If there are important truths to be discovered through introspection, there must be better and worse ways to do this—and one should expect to meet a range of experts, novices, fools, and frauds along the way. Of course, charlatans haunt every walk of life. But on spiritual matters, foolishness and fraudulence can be especially difficult to detect. Unfortunately, this is a natural consequence of the subject matter. When learning to play a sport like golf, you can immediately establish the abilities of the teacher, and the teacher can, in turn, evaluate your progress without leaving anything to the imagination. All the relevant facts are in plain view. If you can't consistently hit the little white ball where you want it to go, you have something to learn from anybody who can. The difference between an expert and a novice is no less stark when it comes to recognizing the illusion of the self. But the qualifications of a teacher and the progress of a student are more difficult to assess.

Spiritual teachers of a certain ability, whether real or imagined, are often described as "gurus," and they elicit an unusual degree of devotion from their students. If your golf instructor were to insist that you shave your head, sleep no more than four hours each night, renounce sex, and subsist on a diet of raw vegetables, you would find a new golf instructor. However, when gurus make demands of this kind, many of their students simply do as directed.

In the West, the term *guru* immediately conjures the image of a surrounding "cult" of devotees—a situation known to give rise to terrifying social deformities. In cults and other fringe spiritual communities, we often find a collection of needy and credulous dropouts ruled by a charismatic psychotic or psychopath. When we consider groups like the People's Temple under Jim Jones, the Branch Davidians under David Koresh, and Heaven's Gate under Marshall Applewhite, it is almost impossible to understand how the spell was first cast, let alone how it was maintained under conditions of such terrible deprivation and danger. But each of these groups proved that intellectual isolation and abuse can lead even well-educated people to willingly destroy themselves.

Gurus fall at every point along the spectrum of moral wisdom. Charles Manson was a guru of sorts. Jesus, the Buddha, Muhammad, Joseph Smith, and every other patriarch and matriarch of the world's religions were as well. For our purpose, the only differences between a cult and a religion are the numbers of adherents and the degree to which they are marginalized by the rest of society. Scientology remains a cult. Mormonism has (just barely) become a religion. Christianity has been a religion for more than a thousand years. But one searches in vain for differences in their respective doctrines that account for the difference in their status.

Some gurus claim to channel the dead, to be poised to leave Earth on an alien spacecraft, or to have once ruled Atlantis. Others impart perfectly reasonable teachings about the nature of the mind and the causes of human suffering—only to make ridiculous claims about cosmology or the origins of disease. To hear that someone is a "guru" tells us almost nothing apart from the fact that some students hold this person in high esteem. Whether their reasons for doing so are good or bad—and whether these people pose a danger to their neighbors—depends upon the content of their beliefs.

Teachers in any field can help or harm their students, and a person's desire to make progress and to win the teacher's approval can often be exploited—emotionally, financially, or sexually. But a guru purports to teach the very art of living, and thus his beliefs potentially encompass every question relevant to the well-being of his students. Apart from parenthood, probably no human relationship offers greater scope for benevolence or abuse than that of guru to disciple. Unsurprisingly, therefore, the ethical failures of the men and women who assume this role can be spectacular and constitute some of the greatest examples of hypocrisy and betrayal to be found anywhere.

The problem of trust is compounded because the line between valid instruction and abuse can be difficult to discern. Given that the entire purpose of a devotee's relationship to a guru is to have his egocentric illusions exposed and undermined, any unwelcome intrusion into his life can potentially be justified as a teaching.

Whenever Gutei Oshō was asked about Zen, he simply raised his finger. Once a visitor asked Gutei's boy attendant, "What does your master teach?" The boy too raised his finger. Hearing of this, Gutei cut off the boy's finger with a knife. The boy, screaming with pain, began to run away. Gutei called to him, and when he turned around, Gutei raised his finger. The boy suddenly became enlightened.[1]

If cutting off a child's finger can count as compassionate instruction, it seems impossible to predict just how fully a spiritual teacher might depart from conventional ethical norms. This is both a theoretical problem in the literature and a psychological one in many spiritual communities: A student's moral intuitions and instincts for self-preservation can always be recast as symptoms of fear and attachment. Consequently, even the most extraordinarily cruel or degrading treatment at the hands of a guru can be interpreted as being for one's own good: *The master wants to have sex with you or your spouse—why would you resist? Can't you see that your impulse to refuse such a generous overture rests on the very illusion of separateness that you want to overcome? Oh, you don't fancy tithing 20 percent of your income to the ashram? Why are you so attached to the fruits of your own labor? What is enlightenment worth to you anyway? You don't like scrubbing toilets and doing yard work for hours at a stretch? Are you above performing such simple acts of service to the Divine? Don't you see that this feeling of self-importance is precisely what must be surrendered before you can recognize your true nature? You found it humiliating when the master had you strip naked and dance in front of your parents and the rest of the congregation? Can't you see that this was just a mirror held up to expose your own egocentricity? Oh, you don't think an enlightened adept would behave this way? Well, what makes you think that your provincial assumptions about enlightenment are true?*

Given the structure of this game, it is little wonder that many people have been harmed by their relationships to spiritual teachers—or that many teachers, given so much power over the lives of others, have abused it. This ethical terrain is all the more confusing because there is no cult leader so deranged or sadistic, or whose fall from grace was so hideous, that one can't find students who will insist that he or she is the Messiah. It is amazing to consider, but there are people still walking this earth who believe that Jim Jones, David Koresh, and Marshall Applewhite were genuine saviors. It is also safe to say that no teacher has been so saintly and impeccable that someone hasn't left his company convinced that he was a dangerous lunatic. If every guru were judged by the worst thing anyone has ever said about him, none would escape hanging.

It is true, however, that the role of guru seems to attract more than its fair share of narcissists and confidence men. Again, this seems to be a natural consequence of the subject matter. One can't fake being an expert gymnast, a rocket scientist, or even a competent cook—at least not for long—but one can fake being an enlightened adept. Those who succeed in doing this are often quite charismatic, because a person can't survive long in this mode unless he can bowl people over. G. I. Gurdjieff set the standard here, and he may have been the first man to return from his travels in the East and establish himself as a proper guru in the West. He was the classic example of a gifted charlatan. He managed to attract a wide following of smart, successful devotees, including the French mathematician Henri Poincaré, the painter Georgia O'Keeffe, and the authors J. B. Priestley, René Daumal, and Katherine Mansfield. He reached other luminaries as well—including Aldous Huxley, T. S. Eliot, and Gerald Heard—through the efforts of his main disciple, P. D. Ouspensky. Frank Lloyd Wright once declared Gurdjieff "the greatest man in the world."[2] Coming from a narcissist of Wright's caliber, this says quite a lot about what sort of impression the man could make.

However, Gurdjieff taught his students that the moon was alive, that it controlled the thoughts and behaviors of unenlightened people, and that it devoured their souls at the moment of death. He used to make visitors to his chateau in Fontainebleau spend long days digging ditches in the sun—only to have them immediately fill them in and begin digging elsewhere. He must have made quite an impression in person, given how long he was able to get away with this mischief. I'm confident that if I were to teach a similarly insane doctrine, all the while demanding painful and pointless sacrifices from my students, I wouldn't have a friend left on earth by the end of the week.

I'm not saying that being forced to do hard and seemingly useless work cannot benefit a person. Consider the Navy SEALs: To become a SEAL, every candidate must pass a qualifying course so arduous that it would constitute torture if imposed on him against his will. This is a selection process that allows the U.S. Navy to produce the most elite special operations force found anywhere. But it is also a *bad* selection process that serves primarily as a rite of passage. It is well known, for instance, that some of the best recruits to the SEAL program are weeded out owing to sheer bad luck. They simply suffer too many injuries to continue with the training or to survive "hell week"— a five-and-a-half-day purgatory of wet sand, dangerous boat drills, calisthenics, hypothermia, and sleeplessness. But those left standing have had an experience of self-overcoming unknown to humanity outside ancient Sparta—and they can be sure that everyone else with whom they will serve in combat has survived the same ordeal.

One of the first things one learns in practicing meditation is that nothing is intrinsically boring—indeed, boredom is simply a lack of attention. Pay sufficient attention, and the mere experience of breathing can reward months and years of steady vigilance. Every guru knows that drudgery can be a way of testing the strength of this insight. And, needless to say, this truth about the human mind can be exploited. The journalist Frances FitzGerald recounts meeting many well-educated disciples of Osho (Bhagwan

Shree Rajneesh)—doctors, lawyers, engineers, professors—doing years of uncompensated menial labor at his Oregon commune.[3] All appeared quite happy with the work, presumably viewing it as an exercise in self-overcoming. Indeed, abandoning one's worldly ambitions to do menial labor—attentively and joyfully—*can* be an exercise in self-overcoming. Here, two truths apparently collide: A person can be exploited and still learn something valuable in the process.

But one must draw the line somewhere, and I think consent should be the governing principle. SEALs in training can drop out at any time, and they are continually encouraged to do so. The inner voice that says they might not have what it takes to be a SEAL is deliberately amplified by their instructors—often by bullhorn—so that those who don't have what it takes will leave the program. That is what distinguishes SEAL training from actual torture. Cults, by contrast, often violate the principle of consent in many ways. I don't deny that a truly enlightened man or woman—that is, one who has fully and permanently unraveled the conventional sense of self—might awaken his or her students by violating certain moral or cultural norms. But extreme examples of such unconventional behavior—often referred to in the literature as "crazy wisdom"—seem to produce the desired result *only* in the literature. Every modern instance of these shenanigans has seemed far more crazy than wise, attesting to nothing so much as the insecurities and sensual desires of the guru in question. Ancient tales of liberating violence, as in the Zen parable above, or of enlightening sexual exploits seem like literary teaching devices, not accurate accounts of how wisdom has been reliably transmitted from master to disciple.

It is usually easy to detect social and psychological problems in any community of spiritual seekers. This seems to be yet another liability inherent to the project of self-transcendence. Many people renounce the world because they can't find a satisfactory place in it, and almost any spiritual teaching can be used to justify a pathological lack of ambition. For someone who has not yet succeeded at anything and who probably fears failure, a doctrine that criticizes the search for worldly success can be very appealing. And devotion to a guru—a combination of love, gratitude, awe, and obedience—can facilitate an unhealthy return to childhood. In fact, the very structure of this relationship can condemn a student to a kind of intellectual and emotional slavery. The writer Peter Marin captured the mood perfectly:

> Obedience to a "perfect master." One could hear, inwardly in them, the gathering of breath for a collective sigh of relief. At last, to be set free, to lay down one's burden, to be a child again—not in renewed innocence, but in restored dependence, in *admitted*, undisguised dependence. To be told, again, what to do, and how to do it. . . . The yearning in the audience was so palpable, their need so thick and obvious, that it was impossible not to feel it, impossible not to empathize with it in some way. Why not, after all? Clearly there are truths and kinds of wisdom to which most persons will not come alone; clearly there are in the world authorities in matters of the spirit, seasoned travelers, guides. Somewhere there must be truths other than the disappointing ones we have; somewhere there must be access to a world larger than this one. And if, to get there, we must put aside all arrogance of will and the stubborn ego, why not? Why not admit what

we do not know and cannot do and submit to someone who both knows and does, who will teach us if we merely put aside all judgment for the moment and obey with trust and goodwill?[4]

A relationship with a guru, or indeed with any expert, tends to run along authoritarian lines. You don't know what you need to know, and the expert presumably does; that's why you are sitting in front of him in the first place. The implied hierarchy is unavoidable. Contemplative expertise exists, and a contemplative expert is someone who can help you realize certain truths about the nature of your own mind.

Unfortunately, the link between self-transcendence and moral behavior is not as straightforward as we might like. It would seem that people can have genuine spiritual insights, and a capacity to provoke those insights in others, while harboring serious moral flaws. It is not always accurate to call such people "frauds": They aren't necessarily *pretending* to have spiritual insights or to be able to produce such experiences in others. But depending on the level of their practice their insights may be an insufficient antidote to the rest of their personalities. The resulting problems can be accentuated by cultural differences. For instance, what is the age of consent for sex? One wouldn't necessarily get the same answers in Bombay and Boston. Certain schools of Buddhism focus on compassion, kindness, and nonharming to an unusual degree, and this offers some protection against abuses of power. But even here one occasionally finds a venerated master with the ethical intuitions of a pirate.

Consider the case of the late Tibetan lama Chögyam Trungpa Rinpoche, who was an inspired teacher but also an occasionally violent drunk and a philanderer. As guru to Allen Ginsberg,

Trungpa attracted many of America's most accomplished poets into his orbit. Once, at a Halloween party for senior students— where W. S. Merwin, the future poet laureate of the United States, and his girlfriend, the poet Dana Naone, were guests—Trungpa ordered his bodyguards to forcibly strip a sixty-year-old woman of her clothing and carry her naked around the meditation hall. This made Merwin and Naone more than a little uncomfortable, and they thought it wise to return to their room for the rest of the night. Noticing their absence, Trungpa asked a group of devotees to find the poets and bring them back to the party. When Merwin and Naone refused to open their door, Trungpa instructed his disciples to break it down. The resulting forced entry led to chaos—wherein Merwin, who was then famous for his pacifism, fought off his attackers with a broken beer bottle, stabbing several in the face and arms. The sight of blood, and his horror over his own actions, apparently collapsed Merwin's defenses, and he and Naone finally allowed themselves to be captured and brought before the guru.

Trungpa, who was by then quite drunk, castigated the pair for their egocentricity and demanded that they take off their clothes. When they refused, he ordered his bodyguards to strip them. By all accounts, Naone became hysterical and begged someone among the crowd of onlookers to call the police. One student attempted to physically intervene. Trungpa himself punched this Samaritan in the face and then ordered his guards to drag the man from the room.

Predictably, many of Trungpa's students viewed the assault on Merwin and Naone as a profound spiritual teaching meant to subdue their egos. Ginsberg, who had not been present at the time, offered the following assessment in an interview: "In the middle of that scene, to yell 'call the police'—do you realize how *vulgar* that was? The Wisdom of the East was being unveiled, and she's going 'call the police!' I mean, shit! Fuck that shit! Strip 'em naked, break down the door!"[5] Apart from having produced a perfect jewel of hippie moral confusion, Ginsberg exposed the riddle at the heart of the traditional guru-devotee relationship. No doubt Merwin and Naone's preference to not dance naked in public had more than a little to do with their attachment to their own privacy and autonomy. And it isn't *inconceivable* that a guru could operate in such a coercive and seemingly unethical way out of a sense of compassion. In fact, it may have been conceivable to Merwin and Naone themselves, even in the aftermath of this humiliating ordeal, because they remained at Trungpa's seminar for several more days to receive further teachings. However, judging from the effect that Trungpa's wild behavior had on both himself (he apparently died from alcoholism) and his students, it is very difficult to view it as the product of enlightened wisdom.

The scandals surrounding Trungpa's organization did not end there. Trungpa had groomed a Western student, Ösel Tendzin, to be his successor. Tendzin was the first Westerner to be honored in this way in any lineage of Tibetan Buddhism. His appointment as "Vajra Regent" had even been approved by the Karmapa, one of the most revered Tibetan masters then living. As it happens, Tendzin was bisexual, highly promiscuous, and rather fond of pressuring his straight male devotees to have sex with him as a form of spiritual initiation. He later contracted HIV but continued to have unprotected sex with more than a hundred men and women without telling them of his condition. Trungpa and several people on the board of his organization knew that the regent was ill and did their best to keep it a secret. Once the scandal broke, Tendzin claimed that Trungpa had promised him that he would do no harm as long as he continued his spiritual practice. Apparently, the virus in his blood didn't care whether he did his spiritual practice or not. At least one of his victims later died of AIDS, having spread HIV to others.

What one encounters in a person like Trungpa is a mind impressively free of shame. This can be a good thing, provided that one happens to also be committed to the well-being of others. But shame serves a crucial social function: It keeps us from behaving like wild animals. Believing in one's own perfect enlightenment is rather like driving a car without brakes—not a problem if you never need to stop or slow down, but otherwise a terrible idea. The belief that he could live beyond conventional moral constraints is explicitly put forward in Trungpa's teaching:

> [Morality] or discipline is not a matter of binding oneself to a fixed set of laws or patterns. For if a bodhisattva is completely selfless, a completely open person, then he will act according to openness, [and] will not have to follow rules; he will simply fall into patterns. It is impossible for the bodhisattva to destroy or harm other people, because he embodies transcendental generosity. He has opened himself completely and so does not discriminate between *this* and *that*. He just acts in accordance with what *is*. . . . If we are completely open, not watching ourselves at all, but being completely open and communicating with situations as they are, then action is pure, absolute, superior. . . . It is an often-used metaphor that the bodhisattva's conduct is like the walk of an elephant. Elephants do not hurry; they just walk slowly and surely through the jungle, one step after another. They just sail right along. They never fall nor do they make mistakes.[6]

The state of freedom and effortless goodwill that Trungpa describes here undoubtedly corresponds to an experience that certain people have and to a perception (whether true or not) that others can form about them. But boundless compassion is one thing; inerrancy is another. The notion that one is incapable of making mistakes poses obvious ethical concerns, no matter what one's level of realization. Anyone who has studied the spread of Eastern spirituality in the West knows that these elephants often stumble—even stampede—injuring themselves and many others in the process.

A person's eyes convey a powerful illusion of inner life. The illusion is true, but it is an illusion all the same. When we look into the eyes of another human being, we seem to see the light of consciousness radiating from the eyes themselves—there is a glint of joy or judgment, perhaps. But every inflection of mood or personality—even the most basic indication that the person is alive—comes not from the eyes but from the surrounding muscles of the face. If a person's eyes look clouded by madness or fatigue, the muscles *orbicularis oculi* are to blame. And if a person appears to radiate the wisdom of the ages, the effect comes not from the eyes but from what he or she is doing with them. Nevertheless, the illusion is a powerful one, and there is no question that the subjective experience of inner radiance can be communicated with the gaze.

It is not an accident, therefore, that gurus often show an unusual commitment to maintaining eye contact. In the best case, this behavior emerges from a genuine comfort in the presence of other people and deep interest in their well-being. Given such a frame of mind, there may simply be no reason to look away. But

maintaining eye contact can also become a way of "acting spiritual" and, therefore, an intrusive affectation. There are also people
who maintain rigid eye lock not from an attitude of openness and
interest or from any attempt to appear open and interested but
as an aggressive and narcissistic show of dominance. Psychopaths
tend to make exceptionally good eye contact.

Whatever the motive behind it, there can be tremendous power
in an unwavering gaze. Most readers will know what I'm talking
about, but if you want to witness a glorious example of the as-
sertive grandiosity that a person's eyes can convey, watch a few
interviews with Osho. I never met Osho, but I have met many
people like him. And the way he plays the game of eye contact is
simply hilarious.[7]

I confess that there was a period in my life, after I first plunged
into matters spiritual, when I became a nuisance in this respect.
Wherever I went, no matter how superficial the exchange, I gazed
into the eyes of everyone I met as though they were my long-lost
lover. No doubt, many people found this more than a bit creepy.
Others considered it a stark provocation. But it also precipitated
exchanges with complete strangers that were fascinating. With
some regularity people of both sexes seemed to become bewitched
by me on the basis of a single conversation. Had I been peddling
some consoling philosophy and been eager to gather students, I
suspect that I could have made a proper mess of things. I defi-
nitely glimpsed the path that many spiritual imposters have taken
throughout history.

Interestingly, when one functions in this mode, one quickly
recognizes all the other people who are playing the same game. I
had many encounters wherein I would meet the eyes of a person
across the room, and suddenly we were playing War of the War-
locks: two strangers holding each other's gaze well past the point

that our primate genes or cultural conditioning would ordinarily countenance. Play this game long enough and you begin to have some very strange encounters.

I don't remember consciously deciding to stop behaving this way, but stop I did. Nevertheless, it is worth paying attention to the type of eye contact one makes. As I already noted, the discomfort one feels when meeting another's gaze seems like nothing more than a ramification of the very feeling of being a self. For this reason, open-eyed meditation with another person can be a very powerful practice. When one overcomes the resistance to staring into another person's eyes, the absence of self-consciousness can be especially vivid.

Eye Contact Meditation

1. Sit across from your partner and simply stare into each other's eyes. (Depending on how far apart you sit, you might have to pick one eye to focus on.)
2 Continue to hold each other's gaze, without speaking.
3. Ignore laughter and other signs of discomfort.

This practice can be combined with the other techniques described in this book, especially mindfulness of breathing and Douglas Harding's inquiry into "headlessness."

Witnessing the misadventures of supposedly enlightened adepts and their devotees can be depressing. But it can also be amusing. I wrote about one such instance in my first book, *The End of Faith*:

I know a group of veteran spiritual seekers who, after searching for a teacher among the caves and dells of the Himalayas for many months, finally discovered a Hindu yogi who seemed qualified to lead them into the ethers. He was as thin as Jesus, as limber as an orangutan, and wore his hair matted, down to his knees. They promptly brought this prodigy to America to instruct them in the ways of spiritual devotion. After a suitable period of acculturation, our ascetic—who was, incidentally, also admired for his physical beauty and for the manner in which he played the drum—decided that sex with the prettiest of his patrons' wives would suit his pedagogical purposes admirably. These relations were commenced at once, and endured for some time by a man whose devotion to wife and guru, it must be said, was now being sorely tested. His wife, if I am not mistaken, was an enthusiastic participant in this "tantric" exercise, for her guru was both "fully enlightened" and as dashing a swain as Lord Krishna. Gradually, this saintly man further refined his spiritual requirements, as well as his appetites. The day soon dawned when he would eat nothing for breakfast but a pint of Häagen-Dazs vanilla ice cream topped with cashews. We might well imagine that the meditations of a cuckold, wandering the frozen-food aisles of a supermarket in search of an enlightened man's enlightened repast, were anything but devotional. This guru was soon sent back to India with his drum.[8]

Ice cream for breakfast. That may tell us everything we need to know. And yet there is no way around the fact that in spiritual matters, as in all others, we must seek instruction from those whom we deem to be more accomplished than ourselves, and the signs of accomplishment are not always clear. With spirituality, the subject matter and the apparent distance between teacher and student seem to create the perfect conditions for self-deception—and thus for misplaced and exploited trust. It is possible, however—with a bit of luck and discrimination—to bypass such problems while still receiving teachings from those who are wiser and more experienced in these matters than oneself.

I offer my own case as a not entirely unusual example. Throughout my twenties, I studied with many teachers who functioned as gurus in the traditional sense, but I never had a relationship with any of them that I find embarrassing in retrospect or that I wouldn't currently recommend to others. I don't know whether to attribute this to good luck or to the fact that there was a line of devotion I was never tempted to cross. Traditionally, one is admonished to view one's guru as perfect. I confess that I could never take this advice seriously—other than in the trivial sense that consciousness itself might be considered perfect in some way, or that a perfect realization of its intrinsic freedom might be possible. Despite how impressive many of my teachers were, they were undoubtedly human and susceptible to the same cultural biases and physical infirmities that define the lives of ordinary people.

For instance, when it came time for Poonja-ji to marry off his niece, he could think of nothing more enlightened than to publish her picture in the singles section of the local newspaper, after having paid a photographer to lighten the color of her skin by several shades. This practice was ubiquitous in India at the time and considered entirely normal. To my eye, however, it was at once deceitful, demeaning, and expressive of bigotry toward people with dark skin. I could only conclude that either enlightenment failed to clear the mind of such cultural residues or Poonja-ji had yet to achieve full enlightenment. In either case, I couldn't view his solution to the problem of marriage as "perfect."

The gurus I have met personally, as well as those whose careers and teachings I have studied at a distance, range from crooks who could be quickly dismissed to teachers who were brilliant but flawed, to those who, while still human, seemed to possess so much compassion and clarity of mind that they were nearly flawless examples of the benefits of spiritual practice. This last group is of obvious interest, and these are surely the people one hopes to meet, but the middle group can be helpful as well. Some teachers about whom depressing stories are told—men and women whose indiscretions may seem to discredit the very concept of spiritual authority—are, in fact, talented contemplatives. Many of these people get corrupted by the power and opportunities that come from inspiring devotion in others. Some may begin to believe the myths that grow up around them, and some are guilty of ludicrous exaggerations of their own spiritual and historical significance. *Caveat emptor.*

Of course, there can be clear indications that a teacher is not worth paying attention to. A history as a fabulist or a con artist should be considered fatal; thus, the spiritual opinions of Joseph Smith, Gurdjieff, and L. Ron Hubbard can be safely ignored. A fetish for numbers is also an ominous sign. Math is magical, but math approached *like* magic is just superstition—and numerology is where the intellect goes to die. Prophecy is also a very strong indication of chicanery or madness on the part of a teacher, and of stupidity among his students. One can extrapolate from scientific data or technological trends (climate models, Moore's law), but most detailed predictions about the future lead to embarrassment right on schedule. Anyone who can confidently tell you what the world will be like in 2027 is delusional. The channeling of invisible entities, whether broadcast from beyond the grave or from another galaxy, should provoke only laughter. J. Z. Knight, who has long claimed to be the mouthpiece for a 35,000-year-old entity named Ramtha, is the ultimate example of how you don't want your teacher to sound. And any suggestion that a guru has influenced world events through magic should also put an end to the conversation. Sri Aurobindo and his partner, known as "the Mother," apparently claimed to have decided the outcome of World War II with their psychic powers.[9] (In that case, one wonders why they weren't held morally responsible for not having ended it sooner.) Yet another reason to ignore Aurobindo's long, unreadable books.

Generally speaking, you should head for the door at any sign of deception on the part of a teacher. Admittedly, you might want to make certain allowances for cultural differences and for the harmlessness of the lie. On one occasion, a very great Dzogchen master—truly one of the most inspiring people I have ever met—declared that a certain day of our retreat would be one of vegetarian austerity (which, from a Tibetan point of view, is an actual sacrifice). Sometime after lunch I entered his room and caught him *in flagrante delicto*, furtively eating a steak out of tinfoil. The moment he saw me, this devilish old lama wadded the foil into a ball and chucked it to his wife like a quarterback delivering a lateral pass. She then hurled it across the room, where it made a distinctly moist thud in the back of a closet. Needless to say, we all had a very good laugh over these machinations, and it was not the sort of deception that seemed calculated to manipulate students or to falsely elevate the status of the teacher. In fact, this teacher did not elevate himself at all—a quality that can compensate for many other sins.

I have never encountered a spiritual teacher who I thought was fully enlightened in the sense that many Buddhists and Hindus imagine is possible—that is, stably free of the illusion of self and endowed with clairvoyance and other miraculous powers. While I remain open to evidence of psi phenomena—clairvoyance, telepathy, and so forth—the fact that they haven't been conclusively demonstrated in the lab is a very strong indication that they do not exist. Researchers who study these things allege that the data are there and that proof of psi can be seen in departures from randomness that occur over thousands of experimental trials.[10] But people who believe that their guru has supernormal powers aren't thinking in terms of weak, statistical effects. They believe that a specific person can reliably read minds, heal the sick, and work other miracles. I have yet to see a case in which evidence for such abilities was presented in a credible way. If one person on earth possessed psychic powers to any significant degree, this would be among the easiest facts to authenticate in a lab. Many people have been duped by traditional evasions on this point; it is often said, for instance, that demonstrating such powers on demand would be spiritually uncouth and that even to want such empirical evidence is an unflattering sign of doubt on the part of a student. *Except ye see signs and wonders, ye will not believe* (John 4:48). A lifetime of foolishness and self-deception awaits anyone who won't call this bluff.

But one need not believe in psychic powers to cut through the illusion of the self. Accomplishing this can be elusive enough. If I've met a person who has done so perfectly, I am unaware of it. I have studied with several people who were assumed to be fully enlightened in that sense, and even some who made the claim explicitly. But as far as I can tell, this added nothing of value to their teachings, while introducing a distracting note of grandiosity into the conversation. Whether or not it's possible for someone to have a permanent experience of self-transcendence, a student's conviction that a teacher is fully enlightened seems superfluous—and it is usually cast in doubt by something silly the teacher says or does in any case.

Once again, I believe that too much can be made of the failures of specific spiritual teachers or of the pathologies found among their followers, as though such pratfalls discredit the guru-disciple relationship in principle. One might draw a useful analogy to marriage here: Examples of bad marriages, or at least unenviable ones, are everywhere to be seen, and few seem to live up to the institution's promise. Focusing on scenes of domestic misery, one might easily conclude that the very idea of marriage is flawed and that human beings should find a better way to arrange themselves and to raise children. I think this conclusion would be reckless. Although I have yet to find a spiritual community that appeared worth joining, and signs of trouble are very easy to spot, I have known many people who learned a great deal by spending extended periods of time in the company of one or another spiritual teacher. And I have learned indispensable things myself.

All this may raise a concern about whether the ideal of enlightenment is a false one. Is true freedom even possible? It certainly is in a momentary sense, as any mature practitioner of meditation knows, and those moments can increase in both number and duration with practice. Therefore, I see no reason why a person couldn't perfectly banish the illusion of the self. However, just the ability to meditate—to rest as consciousness for a few moments prior to the arising of the next thought—can offer a profound relief from mental suffering. We need not come to the end of the path to experience the benefits of walking it.

MIND ON THE BRINK OF DEATH

One cannot travel far in spiritual circles without meeting people who are fascinated by the "near-death experience" (NDE). The phenomenon has been described as follows:

> Frequently recurring features include feelings of peace and joy; a sense of being out of one's body and watching events going on around one's body and, occasionally, at some distant physical location; a cessation of pain; seeing a dark tunnel or void; seeing an unusually bright light, sometimes experienced as a "Being of Light" that radiates love and may speak or otherwise communicate with the person; encountering other beings, often deceased persons whom the experiencer recognizes; experiencing a revival of memories or even a full life review, sometimes accompanied by feelings of judgment; seeing some "other realm," often of great beauty; sensing a barrier or border beyond which the person cannot go; and returning to the body, often reluctantly.[11]

Such accounts have led many people to believe that consciousness must be independent of the brain. However, these experiences vary across cultures, and no single feature is common to them all. One would think that if a nonphysical domain were truly being explored, some universal characteristics would stand out. Hindus and Christians would not substantially disagree—and one certainly wouldn't expect the after-death state of South Indians to diverge from that of North Indians, as has been reported.[12] It should also trouble NDE enthusiasts that only 10 to 20 percent of people who approach clinical death recall having any experience at all.[13]

But the deepest problem with drawing sweeping conclusions from the NDE is that those who have had one and subsequently talked about it *did not die*. Indeed, many of them appear to have been in no actual danger of dying. And those who have reported leaving their bodies during a true medical emergency—after cardiac arrest, for instance—did not suffer a complete loss of brain activity. Even in cases where the brain is alleged to have shut down, its activity must return if the subject is to survive and describe the experience. In such cases, there is generally no way to establish that the NDE occurred while the brain was offline.

Many students of the NDE claim that certain people left their bodies and perceived the commotion surrounding their near death: the efforts of hospital staff to resuscitate them, details of surgery, the grief of family members. Some subjects even say that they learned facts while traveling beyond their bodies that would otherwise have been impossible to know—for instance, a secret told by a dead relative, the truth of which was later confirmed. Reports of this kind seem especially vulnerable to self-deception, if not deliberate fraud. There is another problem, however: Even if true, such phenomena might suggest only that the human mind possesses powers of extrasensory perception (clairvoyance or telepathy, for example). This would be an astonishing discovery, but it wouldn't demonstrate the survival of death. Why? Because unless we could know that a subject's brain was not functioning when the impressions were formed, the involvement of the brain must be presumed.[14]

What is needed to establish the mind's independence from the brain is a case in which a person has an experience—of anything—without associated brain activity. From time to time, someone will claim that a specific NDE meets this criterion. One of the most celebrated cases in the literature involves a woman, Pam Reynolds, who underwent a procedure known as "hypothermic cardiac arrest," in which her core body temperature was brought down to 60 degrees, her heart was stopped, and blood flow to her brain was suspended so that a large aneurysm in her basilar artery could be repaired. Reynolds reports having had a classic NDE, complete with an awareness of the details of her surgery.

Her story presents several problems, however. The events in the world that Reynolds claims to have witnessed during her NDE occurred either before she was "clinically dead" or after blood circulation had been restored to her brain. In other words, despite the extraordinary details of the procedure, we have every reason to believe that Reynolds's brain was functioning when she had her experiences. Also, her case wasn't published until several years after it occurred, and its author, Dr. Michael Sabom, is a born-again Christian who had been working for decades to substantiate the otherworldly significance of the NDE. The possibility that experimenter bias, witness tampering (however unconscious), and false memories intruded into this best of all recorded cases is painfully obvious.

The latest NDE to receive wide acclaim was featured on the cover of *Newsweek* magazine: "Heaven Is Real: A Doctor's Experience of the Afterlife." The great novelty of this case is that its subject, Eben Alexander, is a neurosurgeon who, we might presume, is competent to judge the scientific significance of his experience. Alexander also wrote a book, *Proof of Heaven: A Neurosurgeon's Journey into the Afterlife*, which became an instant bestseller. As it happens, it displaced one of the bestselling books of the past decade, *Heaven Is for Real*, yet another account of the afterlife, based on the near-death adventures of the four-year-old son of a minister.

Unsurprisingly, the two books offer incompatible views of what awaits us beyond the prison of the brain. (Colorful as his account is, Alexander neglects to tell us that Jesus rides a rainbow-colored horse or that the souls of dead children must still do homework in heaven.) At the time of this writing, Alexander's book is ranked #1 on the *New York Times* paperback bestseller list, and it has been on the list for fifty-six weeks. The psychologist Raymond Moody, who coined the phrase "near-death experience," called Alexander's account "the most astounding I have heard in more than four decades of studying this phenomenon. [He] is living proof of an afterlife."[15] Well then, prepare to be astounded.

Once upon a time, a neurosurgeon named Eben Alexander contracted a bad case of bacterial meningitis and fell into a coma. While immobile in his hospital bed, he experienced visions of such intense beauty that they changed everything—not just for him but for all of us, and for science as a whole. According to Alexander, his experience proves that consciousness is independent of the brain, that death is an illusion, and that heaven exists—complete with the usual angels, clouds, and departed relatives but also butterflies and beautiful girls in peasant dress. Our current understanding of the mind "now lies broken at our feet," for, Alexander declares, "what happened to me destroyed it, and I intend to spend the rest of my life investigating the true nature of consciousness and making the fact that we are more, much more, than our physical brains as clear as I can, both to my fellow scientists and to people at large."[16]

As should be clear from the preceding chapters, unlike many scientists and philosophers, I remain agnostic on the question of how consciousness is related to the physical world. There are good reasons to believe that it is an emergent property of brain activity, just as the rest of the human mind is. But we know nothing

about how such a miracle of emergence might occur. And if consciousness were irreducible—or even separable from the brain in a
way that would give comfort to Saint Augustine—my worldview
would not be overturned. I know that we do not understand consciousness, and nothing that I think I know about the cosmos or
about the patent falsity of most religious beliefs requires that I
deny this. So, although I am an atheist who can be expected to be
critical of religious dogma, I am not reflexively hostile to claims of
the sort Alexander has made. In principle, my mind is open. (It
really is.)

However, almost nothing about Alexander's account with-
stands scrutiny—and this is especially insidious, given that he
claims to be a scientist. Many of his errors are glaring but immaterial. In his book, for instance, he understates the estimated
number of neurons in the human brain by a factor of 10. Others
are utterly damning to his case. Whatever his qualifications on
paper, Alexander's evangelizing about his experience in coma is
so devoid of intellectual sobriety, to say nothing of rigor, that I
would see no reason to engage with it—if not for the fact that his
book has been read and believed by millions of people. One of
the greatest obstacles I see to our fashioning a rational approach
to spirituality is to have religious superstition and self-deception
masquerade as science. Hence, it is worth considering Alexander's
case in detail.

First, there are some troubling signs that the good doctor is
just another casualty of American-style Christianity, for though he
claims to have been a nonbeliever before his adventures in coma,
he offers the following self-portrait:

> Although I considered myself a faithful Christian, I was
> so more in name than in actual belief. I didn't begrudge

those who wanted to believe that Jesus was more than simply a good man who had suffered at the hands of the world. I sympathized deeply with those who wanted to believe that there was a God somewhere out there who loved us unconditionally. In fact, I envied such people the security that those beliefs no doubt provided. But as a scientist, I simply knew better than to believe them myself.

What it means to be a "faithful Christian" without "actual belief" is not spelled out, but few nonbelievers will be surprised that our hero's scientific skepticism proves no match for his religious conditioning. Most of us have been around this block often enough to know that many "former atheists," like Francis Collins, spent so long on the brink of faith and yearned for its emotional consolations with such vampiric intensity that the slightest breeze would send them hurtling into the abyss. For Collins, you may recall, all it took to establish the divinity of Jesus and the coming resurrection of the dead was the sight of a frozen waterfall. As we will see, Alexander seems to have required a ride on a psychedelic butterfly. In either case, it's not the perception of beauty we should begrudge but the utter absence of intellectual seriousness with which the author interprets it.

Everything in Alexander's account rests on his repeated and unwarranted assertion that his visions of heaven occurred while his cerebral cortex was "shut down," "inactivated," "completely shut down," "totally offline," and "stunned to complete inactivity." He claims that the cessation of cortical activity was "clear from the severity and duration of my meningitis, and from the global cortical involvement documented by CT scans and neurological examinations." To his editors, this presumably sounded like science.

Unfortunately, the evidence Alexander offers—in the article, in a subsequent response to my public criticism of it, in his book, and in multiple interviews—suggests that he doesn't understand what would constitute compelling evidence for his central claim of cortical inactivity. The proof he offers is either fallacious (CT scans do not measure brain activity) or irrelevant (it does not matter, even slightly, that his form of meningitis was "astronomically rare")—and no combination of fallacy and irrelevancy adds up to sound science. Alexander makes no reference to functional data that might have been acquired by fMRI, PET, or EEG—nor does he seem to realize that this is the sort of evidence necessary to support his case. The impediment to taking Alexander's claims seriously can be simply stated: *There is no reason to believe that his cerebral cortex was inactive at the time he had his experience of the afterlife.* The fact that Alexander thinks he has demonstrated otherwise—by continually emphasizing how sick he was, the infrequency of cases of *E. coli* meningitis, and the ugliness of his initial CT scan—suggests a deliberate disregard of the most plausible interpretation of his experience.

Apparently, Alexander's cortex is functioning now—he has, after all, written a book—so whatever structural damage appeared on CT could not have been "global." Otherwise he would be making the quite crazy claim that his entire cortex was destroyed and then grew back. Coma is not associated with the complete cessation of cortical activity in any case. In fact, neuroimaging studies show that comatose patients (like patients under general anesthesia) have 50 to 70 percent of the normal level of cortical activity.[17] And to my knowledge, almost no one thinks that consciousness is purely a matter of what happens in the cortex.

Why doesn't Alexander know these things? He is, after all, a neurosurgeon who now claims to be upending the scientific worldview on the basis of the fact that his cortex was totally quiescent at the precise moment he was enjoying the best day of his life in the company of angels. Even if his entire cortex did truly shut down (again, an incredible claim), how can he know that his visions didn't occur in the minutes and hours after its functions returned? The very fact that Alexander *remembers* his NDE suggests that the cortical and subcortical structures necessary for memory formation were active at the time. How else could he recall the experience?

Not only does Alexander appear ignorant of the relevant science, he doesn't realize how many people have experienced visions similar to his while under the influence of psychedelics such as DMT or anesthetics such as ketamine. In fact, he has said that any suggestion of similarity between the effect of such compounds on the brain and his experience is "not even in the right ballpark." But here is Alexander's description of the afterlife (as told in an interview):

> I was a speck on a beautiful butterfly wing; millions of other butterflies around us. We were flying through blooming flowers, blossoms on trees, and they were all coming out as we flew through them. . . . [There were] waterfalls, pools of water, indescribable colors, and above there were these arcs of silver and gold light and beautiful hymns coming down from them. Indescribably gorgeous hymns. I later came to call them "angels," those arcs of light in the sky. I think that word is probably fairly accurate. . . .
>
> Then we went out of this universe. I remember just seeing everything receding and initially I felt as if my awareness was in an infinite black void. It was very comforting but I could feel the extent of the infinity and that it was,

as you would expect, impossible to put into words. I was there with that Divine presence that was not anything that I could visibly see and describe, and with a brilliant orb of light. . . .

They said there were many things that they would show me, and they continued to do that. In fact, the whole higher-dimensional multiverse was this incredibly complex corrugated ball and all these lessons [were] coming into me about it. Part of the lessons involved becoming all of what I was being shown. It was indescribable.[18]

"Not even in the right ballpark"? His experience sounds so much like a DMT trip that we are not only in the right ballpark, we are talking about the stitching on the same ball. *Everything* that Alexander describes about his experience, including the parts I have left out, has been reported by DMT users. The similarity is uncanny. Here is how Terence McKenna described the prototypical DMT trance:

Under the influence of DMT, the world becomes an Arabian labyrinth, a palace, a more than possible Martian jewel, vast with motifs that flood the gaping mind with complex and wordless awe. Color and the sense of a reality-unlocking secret nearby pervade the experience. There is a sense of other times, and of one's own infancy, and of wonder, wonder and more wonder. It is an audience with the alien nuncio. In the midst of this experience, apparently at the end of human history, guarding gates that seem surely to open on the howling maelstrom of the unspeakable emptiness between the stars, is the Aeon.

The Aeon, as Heraclitus presciently observed, is a child

at play with colored balls. Many diminutive beings are present there—the tykes, the self-transforming machine elves of hyperspace. Are they the children destined to be father to the man? One has the impression of entering into an ecology of souls that lies beyond the portals of what we naïvely call death. I do not know. Are they the synesthetic embodiment of ourselves as the Other, or of the Other as ourselves? Are they the elves lost to us since the fading of the magic light of childhood? Here is a tremendum barely to be told, an epiphany beyond our wildest dreams. Here is the realm of that which is stranger than we *can* suppose. Here is the mystery, alive, unscathed, still as new for us as when our ancestors lived it fifteen thousand summers ago. The tryptamine entities offer the gift of new language, they sing in pearly voices that rain down as colored petals and flow through the air like hot metal to become toys and such gifts as gods would give their children. The sense of emotional connection is terrifying and intense. The Mysteries revealed are real and if ever fully told will leave no stone upon another in the small world we have gone so ill in.

This is not the mercurial world of the UFO, to be invoked from lonely hilltops; this is not the siren song of lost Atlantis wailing through the trailer courts of crack-crazed America. DMT is not one of our irrational illusions. I believe that what we experience in the presence of DMT is real news. It is a nearby dimension—frightening, transformative, and beyond our powers to imagine, and yet to be explored in the usual way. We must send fearless experts, whatever that may come to mean, to explore and to report on what they find.[19]

Alexander believes that his brain could not have produced his visions because they were too "intense," too "hyper-real," too "beautiful," too "interactive," and too drenched in significance for a brain to conjure. He also thinks that his visions could not have arisen in the minutes or hours during which his cortex (which surely never went off) switched back on. But he has simply ignored what people with working brains experience under the influence of psychedelics. And he does not appear to know that visions of the sort that McKenna describes, although they may seem to last for ages, require only a brief span of biological time. Unlike LSD and other long-acting psychedelics, DMT alters consciousness for only a few minutes. Alexander would have had more than enough time to experience a visionary ecstasy as he was coming out of his coma (whether or not his cortex was rebooting).

Alexander knows that DMT already exists in the brain as a neurotransmitter. Did his brain experience a surge of DMT release during his coma? In his book, he discounts this possibility by reiterating the unfounded claim upon which his entire account rests: DMT would require a functioning cortex upon which to act, whereas his cortex "wasn't available to be affected." Similar experiences can be had with ketamine, a surgical anesthetic that is occasionally used to protect a traumatized brain. Did Alexander by any chance *receive ketamine* while in the hospital? Did he have some other anesthetic that might produce a similar spectrum of effects at low doses? Would he even think it relevant if he had? His assertion that a psychedelic like DMT or an anesthetic like ketamine could not "explain the kind of clarity, the rich interactivity, the layer upon layer of understanding" he experienced is perhaps the most amazing thing he has said since returning from heaven. Such compounds are universally understood to do the job. And most scientists believe that the reliable effects of psychedelics indicate that the brain is at the very least *involved* in the production of visionary states of the sort Alexander is talking about.

The knowledge of the afterlife that Alexander claims to possess also depends upon some extraordinarily dubious methods of verification. While in his coma, he saw a beautiful girl riding beside him on the wing of a butterfly. We learn in his book that he developed his recollection of this experience over a period of *months*—writing, thinking about it, and mining it for new details. It would be hard to imagine a better way to engineer a distortion of memory.

Alexander also tells us that he had a biological sister he never met, who died some years before his coma. Seeing her picture for the first time after his recovery, he judged this woman to be the girl who had joined him for the butterfly ride. He sought further confirmation of this by speaking with his biological family, from whom he learned that his dead sister had, indeed, always been "very loving." QED.

As I've said throughout this book, I have spent much of my life studying and seeking experiences of the kind Alexander describes. I haven't contracted meningitis, thankfully, nor have I had an NDE, but I have experienced many phenomena that often lead people to believe in the supernatural. For instance, I once had an opportunity to study with the great Tibetan lama Dilgo Khyentse Rinpoche in Nepal. Before making the trip, I had a dream in which he seemed to give me teachings about the nature of the mind. The dream struck me as interesting for two reasons: The teachings I received were novel, useful, and convergent with what I later understood to be true, and I had never met Khyentse Rinpoche, nor was I aware of having seen a photograph of him. (This preceded my access to the Internet by at least five years, so the belief that I had never seen his picture was more plausible than it would be now.) I also recall that I had no easy way of finding a picture of him for the sake of comparison. But because I was about to meet the man himself, it seemed that I would be able to confirm whether he had really been in my dream.

First, the teachings: The lama in my dream began by asking who I was. I responded by telling him my name. Apparently, this wasn't the answer he was looking for.

"Who are you?" he said again. He was now staring fixedly into my eyes and pointing at my face with an outstretched finger. I did not know what to say.

"Who are you?" he said again, continuing to point.

"Who are you?" he said a final time, but here he suddenly shifted his gaze and pointing finger, as though he were now addressing someone just to my left. The effect was quite startling, because I knew (insofar as one can be said to know anything in a dream) that we were alone. The lama was pointing to someone who wasn't there, and I suddenly noticed what I would later understand to be an important truth about the nature of the mind: Subjectively speaking, there is only consciousness and its contents; there is no inner self who is conscious. The sense of looking over one's own shoulder, as it were, is an illusion. The lama in my dream seemed to dissect this very feeling of being a self and, for a brief moment, removed it from my mind. I awoke convinced that I had glimpsed something quite profound.

After traveling to Nepal and encountering the arresting figure of Khyentse Rinpoche instructing hundreds of monks from atop a brocade throne, I was struck by the sense that he really did resemble the man in my dream. Even more apparent, however, was the fact that I couldn't know whether this impression was veridical. No doubt, it would have been *more fun* to believe that something magical had occurred and that I had been singled out for some sort of transpersonal initiation—but the allure of this belief suggested only that the bar for proof should be raised rather than lowered. And even though I had no formal scientific training at that point, I knew that human memory is unreliable under conditions of this kind. How much stock could I put in the feeling of familiarity? Was I accurately recalling the face of a man I had met in a dream, or was I engaged in a creative reconstruction of it? If nothing else, the experience of déjà vu proves that one's sense of having experienced something previously can jump the tracks of genuine recollection. My travels in spiritual circles had also brought me into contact with many people who seemed all too eager to deceive themselves about experiences of this kind, and I did not wish to emulate them. Given these considerations, I did not believe that Khyentse Rinpoche had *really* appeared in my dream. And I certainly would never have been tempted to use this experience as conclusive proof of the supernatural.

I invite the reader to compare this attitude to the one that Dr. Eben Alexander will most likely exhibit before crowds of credulous people for the rest of his life. The structure of our experiences was similar: We were each given an opportunity to compare a face remembered from a dream/vision with a person (or photo) in the physical world. I realized that the task was hopeless. Alexander believes that he has made the greatest discovery in the history of science.

Again, nothing can be said against Alexander's experience. And such ecstasies do tell us something about how good a human mind can feel. The problem is that the conclusions Alexander has drawn from his experience—as a *scientist*, he continually reminds us—are based on flagrant errors in reasoning and misunderstandings of the relevant science.

The enthusiastic reception Alexander has enjoyed also suggests a general confusion about the nature of scientific authority. Much of the criticism I have received for dismissing his account focuses on what appear to be his impeccable scientific credentials. However, when debating the validity of evidence and arguments, the point is never that one person's credentials trump another's. Credentials merely offer a rough indication of what a person is likely to know—or should know. If Alexander were drawing reasonable scientific conclusions from his experience, he wouldn't need to be a neuroscientist to be taken seriously; he could be a philosopher—or a coal miner. But he simply isn't thinking like a scientist, and so not even a string of Nobel Prizes would shield him from criticism.[20]

Such is the perennial problem with reports of this kind. Some people are so desperate to interpret the NDE as proof of an afterlife that even those whom one would expect to have a strong commitment to scientific reasoning toss their better judgment out the window. The truth is that, whatever happens after death, it is possible to justify a life of spiritual practice and self-transcendence without pretending to know things we do not know.

THE SPIRITUAL USES OF PHARMACOLOGY

Everything we do is for the purpose of altering consciousness. We form friendships so that we can feel love and avoid loneliness. We eat specific foods to enjoy their fleeting presence on our tongues. We read for the pleasure of thinking another person's thoughts. Every waking moment—and even in our dreams—we struggle to direct the flow of sensation, emotion, and cognition toward states of consciousness that we value.

Drugs are another means toward this end. Some are illegal; some are stigmatized; some are dangerous—though, perversely, these categories only partially intersect. Some drugs of extraordinary power and utility, such as psilocybin (the active compound in "magic mushrooms") and lysergic acid diethylamide (LSD), pose no apparent risk of addiction and are physically well tolerated, and yet one can be sent to prison for their use—whereas drugs such as tobacco and alcohol, which have ruined countless lives, are enjoyed *ad libitum* in almost every society on earth. There are other points on this continuum: MDMA, or Ecstasy, has remarkable therapeutic potential, but it is also susceptible to abuse, and some evidence suggests that it can be neurotoxic.[21]

One of the great responsibilities we have as a society is to educate ourselves, along with the next generation, about which substances are worth ingesting and for what purpose and which are not. The problem, however, is that we refer to all these biologically active materials by a single term, *drugs*, making it nearly impossible to have an intelligent discussion about the psychological, medical, ethical, and legal issues surrounding their use. The poverty of our language has been only slightly eased by the introduction of the term *psychedelics* to differentiate certain visionary compounds, which can produce extraordinary insights, from *narcotics* and other classic agents of stupefaction and abuse.

However, we should not be too quick to feel nostalgia for the counterculture of the 1960s. Yes, crucial breakthroughs were made, socially and psychologically, and drugs were central to the process, but one need only read accounts of the time, such as Joan Didion's *Slouching Towards Bethlehem*, to see the problem with a society bent upon rapture at any cost. For every insight of lasting value produced by drugs, there was an army of zombies with flowers in their hair shuffling toward failure and regret. Turning on, tuning in, and dropping out is wise, or even benign, only if you can then drop into a mode of life that makes ethical and material sense and doesn't leave your children wandering in traffic.

Drug abuse and addiction are very real problems, the remedy for which is education and medical treatment, not incarceration. In fact, the most abused drugs in the United States now appear to be oxycodone and other prescription painkillers. Should these medicines be made illegal? Of course not. But people need to be informed about their hazards, and addicts need treatment. And all drugs—including alcohol, cigarettes, and aspirin—must be kept out of the hands of children.

I discuss issues of drug policy in some detail in my first book, *The End of Faith*, and my thinking on the subject has not changed. The "war on drugs" has been lost and should never have been waged. I can think of no right more fundamental than the right to peacefully steward the contents of one's own consciousness. The fact that we pointlessly ruin the lives of nonviolent drug users by incarcerating them, at enormous expense, constitutes one of the great moral failures of our time. (And the fact that we make room for them in our prisons by paroling murderers, rapists, and child molesters makes one wonder whether civilization isn't simply doomed.)

I have two daughters who will one day take drugs. Of course, I will do everything in my power to see that they choose their drugs wisely, but a life lived entirely without drugs is neither foreseeable nor, I think, desirable. I hope they someday enjoy a morning cup of tea or coffee as much as I do. If they drink alcohol as adults, as they probably will, I will encourage them to do it safely. If they choose to smoke marijuana, I will urge moderation. Tobacco should be shunned, and I will do everything within the bounds of decent parenting to steer them away from it. Needless to say, if I knew that either of my daughters would eventually develop a fondness for methamphetamine or heroin, I might never sleep again. But if they don't try a psychedelic like psilocybin or LSD at least once in their adult lives, I will wonder whether they had missed one of the most important rites of passage a human being can experience.

This is not to say that everyone should take psychedelics. As I will make clear below, these drugs pose certain dangers. Undoubtedly, some people cannot afford to give the anchor of sanity even the slightest tug. It has been many years since I took psychedelics myself, and my abstinence is born of a healthy respect for the risks involved. However, there was a period in my early twenties when I found psilocybin and LSD to be indispensable tools, and some of the most important hours of my life were spent under their influence. Without them, I might never have discovered that there was an inner landscape of mind worth exploring.

There is no getting around the role of luck here. If you are lucky, and you take the right drug, you will know what it is to be enlightened (or to be close enough to persuade you that enlightenment is possible). If you are unlucky, you will know what it is to be clinically insane. While I do not recommend the latter experience, it does increase one's respect for the tenuous condition of sanity, as well as one's compassion for people who suffer from mental illness.

Human beings have ingested plant-based psychedelics for millennia, but scientific research on these compounds did not begin until the 1950s. By 1965, a thousand studies had been published, primarily on psilocybin and LSD, many of which attested to the usefulness of psychedelics in the treatment of clinical depression, obsessive-compulsive disorder, alcohol addiction, and the pain and anxiety associated with terminal cancer. Within a few years, however, this entire field of research was abolished in an effort to stem the spread of these drugs among the public. After a hiatus that lasted an entire generation, scientific research on the pharmacology and therapeutic value of psychedelics has quietly resumed.

Psychedelics such as psilocybin, LSD, DMT, and mescaline all powerfully alter cognition, perception, and mood. Most seem to exert their influence through the serotonin system in the brain, primarily by binding to 5-HT2A receptors (though several have affinity for other receptors as well), leading to increased activity in the prefrontal cortex (PFC). Although the PFC in turn modulates subcortical dopamine production—and certain of these compounds, such as LSD, bind directly to dopamine receptors—the effect of psychedelics appears to take place largely outside dopamine pathways, which could explain why these drugs are not habit-forming.

The efficacy of psychedelics might seem to establish the material basis of mental and spiritual life beyond any doubt, for the introduction of these substances into the brain is the obvious cause of any numinous apocalypse that follows. It is possible, however, if not actually plausible, to seize this evidence from the other end and argue, as Aldous Huxley did in his classic *The Doors of Perception*, that the primary function of the brain may be *eliminative*: Its purpose may be to prevent a transpersonal dimension of mind from flooding consciousness, thereby allowing apes like ourselves to make their way in the world without being dazzled at every step by visionary phenomena that are irrelevant to their physical survival. Huxley thought of the brain as a kind of "reducing valve" for "Mind at Large." In fact, the idea that the brain is a filter rather than the origin of mind goes back at least as far as Henri Bergson and William James. In Huxley's view, this would explain the efficacy of psychedelics: They may simply be a material means of opening the tap.

Huxley was operating under the assumption that psychedelics decrease brain activity. Some recent data have lent support to this view; for instance, a neuroimaging study of psilocybin[22] suggests that the drug primarily reduces activity in the anterior cingulate cortex, a region involved in a wide variety of tasks related to self-monitoring. However, other studies have found that psychedelics increase activity throughout the brain. Whatever the case, the action of these drugs does not rule out dualism, or the existence of realms of mind beyond the brain—but then, nothing does. That is one of the problems with views of this kind: They appear to be unfalsifiable. Physicalism, by contrast, could easily be falsified. If science ever established the existence of ghosts or reincarnation or any other phenomenon that placed the human mind (in whole or in part) outside the brain, physicalism would be dead. The fact that dualists can never say what might count as evidence against their views makes this ancient philosophical position very difficult to distinguish from religious faith.

We have reason to be skeptical of the brain-as-barrier thesis. If the brain were merely a filter on the mind, damaging it should increase cognition. In fact, strategically damaging the brain should be the most reliable method of spiritual practice available to anyone. In almost every case, loss of brain should yield *more mind*. But that is not how the mind works.

Some people try to get around this by suggesting that the brain may function more like a radio, a receiver of conscious states rather than a barrier to them. At first glance, this would appear to account for the deleterious effects of neurological injury and disease, for if one smashes a radio with a hammer, it will no longer function properly. There is a problem with this metaphor, however.

Those who employ it invariably forget that *we are the music, not the radio*. If the brain were nothing more than a receiver of conscious states, it should be impossible to diminish a person's experience of the cosmos by damaging her brain. She might *seem* unconscious from the outside—like a broken radio—but, subjectively speaking, the music would play on.

Specific reductions in brain activity might benefit people in certain ways, unmasking memories or abilities that are being actively inhibited by the regions in question. But there is no reason to think that the pervasive destruction of the central nervous system would leave the mind unaffected (much less improved). Medications that reduce anxiety generally work by increasing the effect of the inhibitory neurotransmitter GABA, thereby diminishing neuronal activity in various parts of the brain. But the fact that dampening arousal in this way can make people feel better does not suggest that they would feel better still if they were drugged into a coma. Similarly, it would be unsurprising if psilocybin reduced brain activity in areas responsible for self-monitoring, because that might, in part, account for the experiences that are often associated with the drug. This does not give us any reason to believe that turning off the brain entirely would yield an increased awareness of spiritual realities.

However, the brain *does* exclude an extraordinary amount of information from consciousness. And, like many who have taken psychedelics, I can attest that these compounds throw open the gates. Positing the existence of a Mind at Large is more tempting in some states of consciousness than in others. But these drugs can also produce mental states that are best viewed as forms of psychosis. As a general matter, I believe we should be very slow to draw conclusions about the nature of the cosmos on the basis of inner experiences—no matter how profound they may seem.

One thing is certain: The mind is vaster and more fluid than our ordinary, waking consciousness suggests. And it is simply impossible to communicate the profundity (or seeming profundity) of psychedelic states to those who have never experienced them. Indeed, it is even difficult to remind *oneself* of the power of these states once they have passed.

Many people wonder about the difference between meditation (and other contemplative practices) and psychedelics. Are these drugs a form of cheating, or are they the only means of authentic awakening? They are neither. All psychoactive drugs modulate the existing neurochemistry of the brain—either by mimicking specific neurotransmitters or by causing the neurotransmitters themselves to be more or less active. Everything that one can experience on a drug is, at some level, an expression of the brain's potential. Hence, whatever one has seen or felt after ingesting LSD is likely to have been seen or felt by someone, somewhere, without it.

However, it cannot be denied that psychedelics are a uniquely potent means of altering consciousness. Teach a person to meditate, pray, chant, or do yoga, and there is no guarantee that anything will happen. Depending upon his aptitude or interest, the only reward for his efforts may be boredom and a sore back. If, however, a person ingests 100 micrograms of LSD, what happens next will depend on a variety of factors, but there is no question that *something* will happen. And boredom is simply not in the cards. Within the hour, the significance of his existence will bear down upon him like an avalanche. As the late Terence McKenna never tired of pointing out, this guarantee of profound effect, for better or worse, is what separates psychedelics from every other method of spiritual inquiry.[23]

Ingesting a powerful dose of a psychedelic drug is like strapping oneself to a rocket without a guidance system. One might wind up somewhere worth going, and, depending on the compound and one's "set and setting," certain trajectories are more likely than others. But however methodically one prepares for the voyage, one can still be hurled into states of mind so painful and confusing as to be indistinguishable from psychosis. Hence the terms *psychotomimetic* and *psychotogenic**** are occasionally applied to these drugs.[24]

I have visited both extremes on the psychedelic continuum. The positive experiences were more sublime than I could ever have imagined or than I can now faithfully recall. These chemicals disclose layers of beauty that art is powerless to capture and for which the beauty of nature itself is a mere simulacrum. It is one thing to be awestruck by the sight of a giant redwood and amazed at the details of its history and underlying biology. It is quite another to spend an apparent eternity in egoless communion with it. Positive psychedelic experiences often reveal how wondrously at ease in the universe a human being can be—and for most of us, normal waking consciousness does not offer so much as a glimmer of those deeper possibilities.

People generally come away from such experiences with a sense that conventional states of consciousness obscure and truncate sacred insights and emotions. If the patriarchs and matriarchs of the world's religions experienced such states of mind, many of their claims about the nature of reality would make *subjective* sense. A beatific vision does not tell you anything about the birth of the cosmos, but it does reveal how utterly transfigured a mind can be by a full collision with the present moment.

However, as the peaks are high, the valleys are deep. My "bad

---

* These terms refer to substances that seem to *mimic* or *cause* the symptoms of psychosis.

trips" were, without question, the most harrowing hours I have ever endured, and they make the notion of hell—as a metaphor if not an actual destination—seem perfectly apt. If nothing else, these excruciating experiences can become a source of compassion. I think it may be impossible to imagine what it is like to suffer from mental illness without having briefly touched its shores.

At both ends of the continuum, time dilates in ways that cannot be described—apart from merely observing that these experiences can seem eternal. I have spent hours, both good and bad, in which any understanding that I had ingested a drug was lost, and all memories of my past along with it. Immersion in the present moment to this degree is synonymous with the feeling that one has always been and will always be in precisely this condition. Depending on the character of one's experience at that point, notions of salvation or damnation may well apply. Blake's line about beholding "Eternity in an hour" neither promises nor threatens too much.

In the beginning, my experiences with psilocybin and LSD were so positive that I did not see how a bad trip could be possible. Notions of "set and setting," admittedly vague, seemed sufficient to account for my good luck. My mental set was exactly as it needed to be—I was a spiritually serious investigator of my own mind—and my setting was generally one of either natural beauty or secure solitude.

I cannot account for why my adventures with psychedelics were uniformly pleasant until they weren't, but once the doors to hell opened, they appeared to have been left permanently ajar. Thereafter, whether or not a trip was good in the aggregate, it generally entailed some excruciating detour on the path to sublimity. Have you ever traveled, beyond all mere metaphors, to the Mountain of Shame and stayed for a thousand years? I do not recommend it.

On my first trip to Nepal, I took a rowboat out on Phewa Lake in Pokhara, which offers a stunning view of the Annapurna range. It was early morning, and I was alone. As the sun rose over the water, I ingested 400 micrograms of LSD. I was twenty years old and had taken the drug at least ten times previously. What could go wrong?

Everything, as it turns out. Well, not everything—I didn't drown. I have a vague memory of drifting ashore and being surrounded by a group of Nepali soldiers. After watching me for a while, as I ogled them over the gunwale like a lunatic, they seemed on the verge of deciding what to do with me. Some polite words of Esperanto and a few mad oar strokes, and I was offshore and into oblivion. I suppose *that* could have ended differently.

But soon there was no lake or mountains or boat—and if I had fallen into the water, I am pretty sure there would have been no one to swim. For the next several hours my mind became a perfect instrument of self-torture. All that remained was a continuous shattering and terror for which I have no words.

An encounter like that takes something out of you. Even if LSD and similar drugs are biologically safe, they have the potential to produce extremely unpleasant and destabilizing experiences. I believe I was positively affected by my good trips, and negatively affected by the bad ones, for weeks and months.

Meditation can open the mind to a similar range of conscious states, but far less haphazardly. If LSD is like being strapped to a rocket, learning to meditate is like gently raising a sail. Yes, it is possible, even with guidance, to wind up someplace terrifying, and some people probably shouldn't spend long periods in intensive practice. But the general effect of meditation training is of settling ever more fully into one's own skin and suffering less there.

As I discussed in *The End of Faith*, I view most psychedelic experiences as potentially misleading. Psychedelics do not guarantee wisdom or a clear recognition of the selfless nature of consciousness. They merely guarantee that the contents of consciousness will change. Such visionary experiences, considered in their totality, appear to me to be ethically neutral. Therefore, it seems that psychedelic ecstasies must be steered toward our personal and collective well-being by some other principle. As Daniel Pinchbeck pointed out in his highly entertaining book *Breaking Open the Head*, the fact that both the Mayans and the Aztecs used psychedelics, while being enthusiastic practitioners of human sacrifice, makes any idealistic connection between plant-based shamanism and an enlightened society seem terribly naïve.

The form of transcendence that appears to link directly to ethical behavior and human well-being is that which occurs in the midst of ordinary waking life. It is by ceasing to cling to the contents of consciousness—to our thoughts, moods, and desires—that we make progress. This project does not in principle require that we experience *more* content. The freedom from self that is both the goal and the foundation of spiritual life is coincident with normal perception and cognition—though, as I have already said, this can be difficult to realize.[25]

The power of psychedelics, however, is that they often reveal, in the span of a few hours, depths of awe and understanding that can otherwise elude us for a lifetime. William James said it about as well as anyone:[26]

> One conclusion was forced upon my mind at that time, and my impression of its truth has ever since remained unshaken. It is that our normal waking consciousness, rational consciousness as we call it, is but one special type

of consciousness, whilst all about it, parted from it by the filmiest of screens, there lie potential forms of consciousness entirely different. We may go through life without suspecting their existence; but apply the requisite stimulus, and at a touch they are there in all their completeness, definite types of mentality which probably somewhere have their field of application and adaptation. No account of the universe in its totality can be final which leaves these other forms of consciousness quite disregarded. How to regard them is the question—for they are so discontinuous with ordinary consciousness. Yet they may determine attitudes though they cannot furnish formulas, and open a region though they fail to give a map. At any rate, they forbid a premature closing of our accounts with reality.[27]

I believe that psychedelics may be indispensable for some people—especially those who, like me, initially need convincing that profound changes in consciousness are possible. After that, it seems wise to find ways of practicing that do not present the same risks. Happily, such methods are widely available.

This chapter has taken us along the edge of a precipice. There is no question that novel and intense experiences—whether had in the company of a guru, on the threshold of death, or by recourse to certain drugs—can send one spinning into delusion. But they can also broaden one's view.

The aims of spirituality are not exactly those of science, but neither are they unscientific. Search your mind, or pay attention to the conversations you have with other people, and you will discover that there are no real boundaries between science and any other discipline that attempts to make valid claims about the world on the basis of evidence and logic. When such claims and their methods of verification admit of experiment and/or mathematical description, we tend to say that our concerns are "scientific"; when they relate to matters more abstract, or to the consistency of our thinking itself, we often say that we are being "philosophical"; when we merely want to know how people behaved in the past, we dub our interests "historical" or "journalistic"; and when a person's commitment to evidence and logic grows dangerously thin or simply snaps under the burden of fear, wishful thinking, tribalism, or ecstasy, we recognize that he is being "religious."

The boundaries between true intellectual disciplines are currently enforced by little more than university budgets and architecture. Is the Shroud of Turin a medieval forgery? This is a question of history, of course, and of archaeology, but the techniques of radiocarbon dating make it a question of chemistry and physics as well. The real distinction we should care about—the observation of which is the sine qua non of the scientific attitude—is between demanding good reasons for what one believes and being satisfied with bad ones. Spirituality requires the same commitment to intellectual honesty.

Once one recognizes the selflessness of consciousness, the practice of meditation becomes just a means of getting more familiar with it. The goal, thereafter, is to cease to overlook what is already the case. Paradoxically, this still requires discipline, and setting aside time for meditation is indispensable. But the true discipline is to remain committed, throughout the whole of one's life, to waking up from the dream of the self. We need not take anything on faith to do this. In fact, the only alternative is to remain confused about the nature of our minds.

Consciousness is the basis of both the examined and the unexamined life. It is all that can be seen and that which does the seeing. No matter how far you have traveled from the place of your birth, and however much you now understand about the world, you have been exploring consciousness and its changes. Why not do so directly?

Conclusion

Sometime around her third birthday, my daughter asked, "Where does gravity come from?" After talking about how objects attract each other—and wisely ignoring the curvature of space-time—my wife and I arrived at our deepest and most honest answer: "We don't know. Gravity is a mystery. People are still trying to figure it out."

This type of answer continues to divide humanity. We could have said, as billions of people would have, "Gravity comes from God." But this would have merely stifled our daughter's intelligence—and taught her to stifle it. We could have told her, "Gravity might be God's way of dragging people to hell, where they burn in fire. And you will burn there *forever* if you doubt that God exists." No Christian or Muslim can offer a compelling reason why we shouldn't have said such a thing—or the moral equivalent—and yet that would have been nothing less than the emotional and intellectual abuse of a child. I have now heard from many thousands of people who were oppressed in this way, from the moment they could speak, by the terrifying ignorance and fanaticism of their parents. The reason for this widespread mistreatment of children is clear: Most people still believe that religion provides something essential that cannot be had any other way.

Twelve years have now passed since I first realized how high the stakes are in this war of ideas. I remember feeling the jolt of history when the second plane crashed into the World Trade Center. For many of us, that was the moment we understood that things can go terribly wrong in our world—not because life is unfair or moral progress impossible but because we have failed, generation after generation, to abolish the delusions and animosities of our ignorant ancestors. The worst ideas continue to thrive—and are still imparted, in their purest form, to children.

What is the meaning of life? What is our purpose on earth? These are some of the great, false questions of religion. We need not answer them, for they are badly posed, but we can live our answers all the same. At a minimum, we can create the conditions for human flourishing in this life—the only life of which any of us can be certain. That means we should not terrify our children with thoughts of hell or poison them with hatred for infidels. We should not teach our sons to consider women their future property or convince our daughters that they are property even now. And we must decline to tell our children that human history began with bloody magic and will end with bloody magic in a glorious war between the righteous and the rest.

Such sins against reason and compassion do not represent the totality of religion, but they lie at its core. As for the rest—charity, community, ritual, and the contemplative life—we need not take anything on faith to embrace those goods. It is one of the most damaging lies of religion—whether liberal, moderate, or extreme—to insist that we must.

Spirituality remains the great hole in secularism, humanism, rationalism, atheism, and all the other defensive postures that reasonable men and women strike in the presence of unreasonable faith. People on both sides of this divide imagine that visionary experience has no place within the context of science—apart from the corridors of a mental hospital. Until we can talk about spirituality in rational terms—acknowledging the validity of self-transcendence—our world will remain shattered by dogmatism. This book has been my attempt to begin such a conversation.

There is experience, and then there are the stories we tell about it. At its best, religion is a set of stories that recount the ethical and contemplative insights of our wisest ancestors. But these stories come to us bundled with ancient confusion and perennial lies. And they invariably harden into doctrines that defy revision, generation after generation. The great pressure of accumulating knowledge—in science, medicine, history—has begun to scour our culture of many of these ideas. With the force of a glacier, perhaps, but at a similar pace. The exponential increase in the power of technology brings with it a commensurate increase in the consequences of human ignorance. We do not have centuries to wait for our neighbors to come to their senses.

Religious stories may bring meaning to people's lives, but some meanings are patently false and divisive. What does a spiritual experience mean? If you are a Christian sitting in church, it might mean that Jesus Christ survived his death and has taken a personal interest in the fate of your soul. If you are a Hindu praying to Shiva, you will have a very different story to tell. Altered states of consciousness are empirical facts, and human beings experience them under a wide range of conditions. To understand this, and to seek to live a spiritual life without deluding ourselves, we must view these experiences in universal and secular terms.

Happiness and suffering, however extreme, are mental events. The mind depends upon the body, and the body upon the world, but everything good or bad that happens in your life must appear in consciousness to matter. This fact offers ample opportunity to make the best of bad situations—changing your perception of the world is often as good as changing the world—but it also allows a person to be miserable even when all the material and social conditions for happiness have been met. During the normal course of events, your mind will determine the quality of your life.

Of course, the mind is as contingent as the body—and the limits of the body are obvious: I am precisely as tall as I am, and not an inch taller. I can jump as high as I can, and no higher. I can't see what is behind my head. My knee hurts. The boundaries of my mind are just as clear: I cannot speak a word of Korean. I don't remember what I did on this date in 2011, or the last words I read of Dante, or even the first words I spoke to my wife this morning. Although I can alter my mood and states of attention, I can do so only within a narrow range. If I am tired, I can open my eyes a little wider and try to perk myself up, but I cannot completely banish the feeling of fatigue. If I am slightly depressed, I can brighten my mood with happy thoughts. I can even access a feeling of happiness directly by simply recalling what it is like to be happy—deliberately putting a smile in my mind—but I cannot reproduce the greatest joy I have ever felt. Everything about my mind and body seems to feel the weight of the past. I am just as I am.

But consciousness is different. It appears to have no form at all, because anything that would give it form must arise *within* the field of consciousness. Consciousness is simply the light by which the contours of mind and body are known. It is that which is aware of feelings such as joy, regret, amusement, and despair. It can seem to take their shape for a time, but it is possible to recognize that it never quite does. In fact, we can directly experience that consciousness is never improved or harmed by what it knows. Making this discovery, again and again, is the basis of spiritual life.

As we have seen, there is no compelling reason to believe that the mind is independent of the brain. And yet the deflationary attitude toward consciousness taken by many scientists—wherein reality is considered only from the outside, in third-person terms—is also unwarranted. A middle path exists between making religion out of spiritual life and having no spiritual life at all.

We have long known that how things seem in the world can be misleading, and this is no less true of the mind itself. And yet many people have found that through sustained introspection, how things seem can be brought into closer register with how they are. In one sense, the science that underlies this claim is in its infancy—but in another, it is complete. Although we are only beginning to understand the human mind at the level of the brain, and we know nothing about how consciousness itself comes into being, it isn't too soon to say that the conventional self is an illusion. There is no place for a soul inside your head. Consciousness itself is divisible—as we saw in the case of split-brain patients—and even in an intact brain consciousness is blind to most of what the mind is doing. Everything we take ourselves to be at the level of our subjectivity—our memories and emotions, our capacity for language, the very thoughts and impulses that give rise to our behavior—depends upon distinct processes that are spread out over the whole of the brain. Many of these can be independently interrupted or extinguished. The sense, therefore, that we are unified subjects—the unchanging thinkers of thoughts and experiencers of experience—is an illusion. The conventional self is a transitory appearance among transitory appearances, and it vanishes when looked for. We need not await any data from the lab to say that self-transcendence is possible. And we need not become masters of meditation to realize its benefits. It is within our capacity to recognize the nature of thoughts, to awaken from the dream of being merely ourselves and, in this way, to become better able to contribute to the well-being of others.

Spirituality begins with a reverence for the ordinary that can lead us to insights and experiences that are anything but ordinary. And the conventional opposition between humility and hubris has no place here. Yes, the cosmos is vast and appears indifferent to our mortal schemes, but every present moment of consciousness is profound. In subjective terms, each of us is identical to the very principle that brings value to the universe. Experiencing this directly—not merely thinking about it—is the true beginning of spiritual life.

We are always and everywhere in the presence of reality. Indeed, the human mind is the most complex and subtle expression of reality we have thus far encountered. This should grant profundity to the humble project of noticing what it is like to be you in the present. However numerous your faults, something in you at this moment is pristine—and only you can recognize it.

Open your eyes and see.

ACKNOWLEDGMENTS

I am especially indebted to my friends Jeff Forrester, Joseph Goldstein, Daniel Goleman, and D.A. Wallach, who read the manuscript of *Waking Up* and offered helpful feedback and encouragement. Andres Fossas provided valuable research assistance and insightful notes on the text. And my copyeditor, Martha Spaulding, did much to improve the clarity of the writing throughout.

Parts of *Waking Up* were derived from the dissertation I wrote in the Interdepartmental PhD Program for Neuroscience at UCLA. Those sections benefited from the guidance of my thesis committee: Mark Cohen, Marco Iacoboni, Eran Zaidel, and Jerome ("Pete") Engel. Paul Churchland, Daniel Dennett, Owen Flanagan, and Steven Pinker also reviewed that material at an early stage and gave very useful comments.

I began writing *Waking Up* just as the publishing industry was entering a period of upheaval. It wasn't long, in fact, before every person I knew at Free Press disappeared in a major round of consolidation at the parent company, Simon & Schuster. At least three people from the previous regime deserve my thanks: Martha Levin, Dominick Anfuso, and Hilary Redmon. I continue to profit from their initial enthusiasm for the project.

Thomas LeBien inherited *Waking Up* at Simon & Schuster and proved to be an outstanding editor. It has been a true pleasure working with him at each stage of the publishing process.

I am also grateful for the continued assistance of my agents, John Brockman, Katinka Matson, and Max Brockman.

For all my books, my mother's notes are among the most important, but for *Waking Up* I owe her a special debt: The time I spent in India and Nepal in my twenties—and in silence at various meditation centers around the world—was made possible by her support. She also gave me my love of books, so it is always a special pleasure to deliver her another one.

As described in the text, I was privileged to learn from some remarkable masters of meditation: Tulku Urgyen Rinpoche, Nyoshul Khen Rinpoche, H. W. L. Poonja, and Sayadaw U Pandita each provided a crucial piece of the puzzle. I am indebted to Joseph Goldstein and Sharon Salzberg, too, for their years of friendship and for the many months I practiced under their roof at the Insight Meditation Society in Barre, Massachusetts.

Finally, my greatest thanks go to Annaka, my wife and best-possible friend. Writing is among the more isolating professions, and I am extraordinarily lucky that the woman I love is also an editor and collaborator on all my projects. When she isn't busy raising our girls to be compassionate human beings—or silencing the howls of displeasure that sometimes emanate from my office—Annaka inspires my thoughts to turn in new directions and improves them on the page. I couldn't do what I do without her.

NOTES

Chapter 1: Spirituality

1. My late friend Christopher Hitchens—no enemy of the lexicographer—didn't share them either. Hitch believed that *spiritual* was a term we could not do without. It is true that he didn't think about spirituality in precisely the way I do. He spoke instead of the spiritual pleasures afforded by certain works of poetry, music, and art. The symmetry and beauty of the Parthenon embodied this happy extreme for him—without there being any need to admit the existence of the goddess Athena, much less devote ourselves to her worship. Hitch also used the terms *numinous* and *transcendent* to mark occasions of great beauty or significance, and for him the Hubble Deep Field was an example of both. (I'm sure he was aware that pedantic excursions into the *OED* would produce etymological embarrassments regarding these words as well.) Carl Sagan also freely used the term *spiritual* in this way. (See C. Sagan. 1995. *The Demon-Haunted World*. New York: Random House. p. 29.)

   I have no quarrel with Hitch and Sagan's general use of *spiritual* to mean something like "beauty or significance that provokes awe," but I believe that we can also use it in a narrower and, indeed, more personally transformative sense.

2. A. Huxley. [1945] 2009. *The Perennial Philosophy: An Interpretation of the Great Mystics, East and West*. New York: Harper Perennial, p. vii.

3. One can speak about Judaism without its myths and miracles—even without God—but this doesn't make Judaism the equivalent of Buddhism. Buddhism without the unjustified bits is essentially a first-person science. Secular Judaism isn't.

4. A. Rawlinson. 1997. *The Book of Enlightened Masters*. Chicago: Open Court, p. 38.

5. For an entertaining account of Blavatsky's career, see P. Washington. 1993. *Madame Blavatsky's Baboon*. New York: Schocken.

6. One wonders how it was possible for a charlatan like L. Ron Hubbard to acquire any following at all, because each story about him is more preposterous and embarrassing than the last. For instance, Hubbard claimed to have withdrawn one of his first books from publication " 'because the first six people who read it were so shattered by the revelations that they had lost their minds' " (L. Wright. 2013. *Going Clear: Scientology, Hollywood, and the Prison of Belief*. New York: Knopf). According to Hubbard, when he delivered this "dangerous text to his publisher, 'The reader brought the manuscript into the room, set it on the publisher's desk, then jumped out the window of the skyscraper.' "

   There are many more laughs to be had at Hubbard's expense. However, several readers who saw the original version of this endnote found it so funny that they had to be hospitalized. Regrettably, I've been forced to edit the text out of concern for the health of my readers.

7. A. Koestler. 1960. *The Lotus and the Robot*. New York: Harper & Row, p. 285. Koestler was also less than impressed with the spiritual efficacy of psychedelics. See A. Koestler. 1968. "Return Trip to Nirvana." In *Drinkers of Infinity: Essays 1955–1967*. London: Hutchinson, pp. 201–12.

8. C. Hitchens. 1998. "His Material Highness." Salon.com.

9. Purists will insist on important differences among the various schools of Buddhism and between Buddhism and the tradition of Advaita Vedanta developed by Shankara (788–820). Although I touch upon some of these distinctions, I do not make much of them. I consider the differences to be generally a matter of emphasis, semantics, and (irrelevant) metaphysics—and too esoteric to be of interest to the general reader.

10. The research on pathological responses to meditation is quite sparse. Traditionally, it is believed that certain stages on the contemplative path are by nature unpleasant and that some forms of mental pain should therefore be considered signs of progress. It

seems clear, however, that meditation can also precipitate or unmask psychological illness. As with many other endeavors, distinguishing help from harm in each instance can be difficult. As far as I know, Willoughby Britton is the first scientist to study this problem systematically.

11. Consider the sensation of touching your finger to your nose. We experience the contact as simultaneous, but we know that it can't be simultaneous at the level of the brain, because it takes longer for the nerve impulse to travel to sensory cortex from your fingertip than it does from your nose—and this is true no matter how short your arms or long your nose. Our brains correct for this discrepancy in timing by holding these inputs in memory and then delivering the result to consciousness. Thus, your experience of the present moment is the product of layered memories.

12. F. Zeidan et al. 2011. "Brain Mechanisms Supporting the Modulation of Pain by Mindfulness Meditation." *Pain* 31: 5540–48; B. K. Holzel et al. 2011. "How Does Mindfulness Meditation Work? Proposing Mechanisms of Action from a Conceptual and Neural Perspective." *Perspectives on Psychological Science* 6: 537–59; B. Kim et al. 2010. "Effectiveness of a Mindfulness-Based Cognitive Therapy Program as an Adjunct to Pharmacotherapy in Patients with Panic Disorder." *J Anxiety Disord* 24(6): 590–95; K. A. Godfrin and C. van Heeringen. 2010. "The Effects of Mindfulness-Based Cognitive Therapy on Recurrence of Depressive Episodes, Mental Health and Quality of Life: A Randomized Controlled Study." *Behav Res Ther* 48(8): 738–46; F. Zeidan, S. K. Johnson, B. J. Diamond, Z. David, and P. Goolkasian. 2010. "Mindfulness Meditation Improves Cognition: Evidence of Brief Mental Training." *Conscious Cogn* 19(2): 597–605; B. K. Hölzel et al. 2011. "Mindfulness Practice Leads to Increases in Regional Brain Gray Matter Density." *Psychiatry Res* 191(1): 36–43.

13. Nanamoli, orig. trans., and Bodhi, trans. and ed. 1995. *The Middle Length Discourses of the Buddha: A New Translation of the Majjhima Nikaya.* Boston: Wisdom Publications.

14. However one bounds the concept of enlightenment, there is no escaping the fact that most traditional accounts of it, Buddhist and otherwise, attribute a variety of supernormal powers to spiri-

tual adepts. Is there any evidence that human beings can acquire abilities like clairvoyance and telekinesis? Apart from anecdotes offered by people who are desperate to believe in such powers, we can say that the evidence is impressively thin. Traditionally, gurus and their devotees have sought to have it both ways: The guru will display various *siddhi*s (Sanskrit: "powers") to entertain and persuade the faithful—but never in such a way as to meet the tests of true skeptics. We are invariably told that to produce miracles on demand would be a crude misuse of a guru's office. The *dharma* (Sanskrit: "way" or "truth"), after all, is more precious and profound than worldly powers. No doubt it is. But this doesn't stop most gurus from taking credit, or their devotees from bestowing it, whenever random coincidences occur.

15. M. Ricard. 2007. *Happiness: A Guide to Developing Life's Most Important Skill.* New York: Little, Brown, p. 19.

Chapter 2: The Mystery of Consciousness

1. T. Nagel. 1974. "What Is It Like to Be a Bat?" *Philosophical Review* 83.
2. One could argue that this notion of "trading places" is fraught with confusion, but Nagel's notion of consciousness being identical to subjective experience isn't.
3. It's true that some philosophers and neuroscientists will want to pull the brakes right here. Daniel Dennett, with whom I agree about many things, tells me that if I can't imagine the falsehood of a statement like "Either the lights are on, or they are not," I'm not trying hard enough. However, on a question as rudimentary as the ontology of consciousness, the debate often comes down to irreconcilable intuitions. While I will try my best to unpack my intuition that the above statement cannot be false, at a certain point a person has to admit that he can't understand what his opponents are talking about.
4. The picture does not change (much) if you are a dualist who believes that brains are conscious only because consciousness is somehow inserted into them. There are many problems with dualism, but even a dualist should agree that consciousness appears to be associated only with organisms of sufficient complexity. Whether

or not one is a dualist, one has no compelling reason to believe that there is something that it is like to be a tomato.

5. Saying that a creature is conscious, therefore, is not to make a claim about its behavior or its use of language, because we can find examples of both behavior and language without consciousness (a primitive robot) and consciousness without either (a person suffering "locked-in syndrome"). Of course, it is possible that some robots are conscious—and if consciousness is the sort of thing that comes into being purely by virtue of information processing, then our cell phones and coffeemakers may be conscious. But few of us imagine that there is something that it is like to be even the most advanced computer. Whatever its relationship to information processing, consciousness is an internal reality that cannot necessarily be appreciated from the outside and need not be associated with behavior or responsiveness to stimuli. If you doubt this, read *The Diving Bell and the Butterfly* (1997), Jean Dominique-Bauby's astonishing and heartbreaking account of his own "locked-in syndrome," which he dictated by signing to a nurse with his left eyelid. Then try to imagine his predicament if even this degree of motor control had been denied him.

6. Descartes is probably the first Western philosopher to make this point, but others have continued to emphasize it, notably the philosophers John Searle and David Chalmers. I do not agree with Descartes's dualism or with some of what Searle and Chalmers have said about the nature of consciousness, but I agree that its subjective reality is both primary and indisputable. This does not rule out the possibility that consciousness is, in fact, identical to certain brain processes.

Again, I should say that some philosophers, such as Daniel Dennett and Paul Churchland, just don't buy this. But I do not understand why. My not seeing how consciousness can possibly be an illusion entails my not understanding how they (or anyone else) can think that it might be one. I agree that we may be profoundly mistaken about consciousness—about how it arises, about its connection to the brain, about precisely what we are conscious of and when. But this is not the same as saying that consciousness itself may be illusory. The state of being completely confused about the nature of consciousness is itself a demonstration of consciousness.

7. "The stuff of the world is mind-stuff." A. S. Eddington. 1928. *The Nature of the Physical World.* Cambridge, UK: Cambridge University Press, p. 276.

    "The old dualism of mind and matter . . . seems likely to disappear . . . through substantial matter resolving itself into a creation and manifestation of mind." J. Jeans. 1930. *The Mysterious Universe.* Cambridge, UK: Cambridge University Press, p. 158.

    "The only acceptable point of view appears to be the one that recognizes both sides of reality—the quantitative and the qualitative, the physical and the psychical—as compatible with each other, and can embrace them simultaneously." W. Pauli, C. P. Enz, and K. v. Meyenn. [1955] 1994. *Writings on Physics and Philosophy.* New York: Springer-Verlag, p. 259.

    "The conception of the objective reality of the elementary particles has thus evaporated not into the cloud of some obscure new reality concept, but into the transparent clarity of a mathematics that represents no longer the behavior of the particle but rather our knowledge of this behavior." W. Heisenberg. 1958. "The Representation of Nature in Contemporary Physics." *Daedalus* 87 (Summer): 100.

    "We simply cannot see how material events can be transformed into sensation and thought, however many textbooks . . . go on talking nonsense on the subject." E. Schrödinger. 1964. *My View of the World,* trans. C. Hastings. Cambridge, UK: Cambridge University Press, pp. 61–62.

8. F. Dyson. 2002. "The Conscience of Physics." *Nature* 420 (December 12): 607–8.

9. I am grateful to my friend, physicist Lawrence Krauss, for clarifying several of these points.

10. If we look for consciousness in the physical world, we find only complex systems giving rise to complex behavior—which may or may not be attended by consciousness. The fact that the behavior of our fellow human beings persuades us that they are conscious (more or less) does not get us any closer to linking consciousness to physical events. Is a starfish conscious? It seems clear that we

will not make any progress on this question by drawing analogies between starfish behavior and our own. Only in the presence of animals sufficiently like ourselves do our intuitions about (and attributions of) consciousness begin to crystallize. Is there something that it is like to be a cocker spaniel? Does it feel its pains and pleasures? Surely it must. How do we know? Behavior and analogy.

Some scientists and philosophers have formed the mistaken impression that it is always more parsimonious to deny consciousness in lower animals than to attribute it to them. I have argued elsewhere that this is not the case (S. Harris. 2004. *The End of Faith: Religion, Terror, and the Future of Reason.* New York: Norton, pp. 276–77). To deny consciousness in chimpanzees, for instance, is to assume the burden of explaining *why* their genetic, neuroanatomical, and behavioral similarity to us is an insufficient basis for it. (Good luck.)

11. The idea that consciousness is identical to (or emerged from) a certain class of unconscious physical events seems impossible to properly conceive—which is to say that we can think we are thinking it, but we are probably mistaken. We can say the right words: "Consciousness emerges from unconscious information processing." We can also say "Some squares are as round as circles" and "2 plus 2 equals 7." But are we really thinking these things all the way through? I don't think so.

12. J. Levine. 1983. "Materialism and Qualia: The Explanatory Gap." *Pacific Philosophical Quarterly* 64.

13. D. J. Chalmers. 1996. *The Conscious Mind: In Search of a Fundamental Theory.* New York: Oxford University Press.

14. This maneuver has its antecedents in the "neutral monism" (so dubbed by Russell) of James and Mach. It is a view I substantially agree with. Here is Nagel on the subject:

> What will be the point of view, so to speak, of such a theory? If we could arrive at it, it would render transparent the relation between mental and physical, not directly, but through the transparency of their common relation to something that is not merely either of them. Neither the mental nor the physical point of view will do for this purpose. The mental will not do because it simply leaves out the physiology, and

has no room for it. The physical will not do because while it includes the behavioral and functional manifestations of the mental, this doesn't, in view of the falsity of conceptual reductionism, enable it to reach to the mental concepts themselves. . . . The difficulty is that such a viewpoint cannot be constructed by the mere conjunction of the mental and the physical. It has to be something genuinely new, otherwise it will not possess the necessary unity. . . . Such a conception will have to be created; we won't just find it lying around. All the great reductive successes in the history of science have depended on theoretical concepts, not natural ones—concepts whose whole justification is that they permit us to replace brute correlations with reductive explanations. At present such a solution to the mind-body problem is literally unimaginable, but it may not be impossible." (T. Nagel. 1998. "Conceiving the Impossible and the Mind-Body Problem." *Philosophy* 73[285]: pp. 337–52.)

15. J. R. Searle. 1992. *The Rediscovery of the Mind.* Cambridge, MA: MIT Press, 1992; J. R. Searle. 2007. "Dualism Revisited." *J Physiol Paris* 101 (4–6); J. R. Searle. 1998. "How to Study Consciousness Scientifically." *Philos Trans R Soc Lond B Biol Sci* 353 (1377).

16. J. Kim. 1993. "The Myth of Nonreductive Materialism." In *Supervenience and Mind.* Cambridge, UK: Cambridge University Press.

17. C. McGinn. 1989. "Can We Solve the Mind-Body Problem?" *Mind* 98; C. McGinn. 1999. *The Mysterious Flame: Conscious Minds in a Material World.* New York: Basic Books. Steven Pinker also throws his lot in with McGinn: S. Pinker. 1997. *How the Mind Works.* New York: Norton, pp. 558–65. This is more or less where Thomas Nagel comes out, though he considers himself less pessimistic than McGinn: Nagel, "Conceiving the Impossible and the Mind-Body Problem."

18. Whatever its relation to the physical world, consciousness seems to be conceptually irreducible, because any attempt to define *consciousness* or its surrogates (*sentience, awareness, subjectivity*) leads us in a lexical circle. One of the great obstacles to understanding consciousness probably lurks here: If an adequate, noncircu-

lar definition of consciousness exists, no one has found it. The same can be said about any idea that is truly basic to our thinking. The reader is invited to try to define the word *causation* in noncircular terms. Consequently, many philosophers and scientists change the subject whenever the discussion turns to matters of consciousness—conflating it with attention, self-awareness, wakefulness, responsiveness to stimuli, or some other, more tractable and less fundamental aspect of cognition. These digressions are often inadvertent and rarely aim at a reductive definition of "consciousness." Where they do, as in the case of (analytical) behaviorism, they invariably seem false and question-begging.

19. Be it "40-Hz coherent activity in thalamocortical pathways" (R. Llinas. 2001. *I of the Vortex: From Neurons to Self.* Cambridge, MA: MIT Press; R. Llinas et al. 1998. "The Neuronal Basis for Consciousness." *Philos Trans R Soc Lond B Biol Sci* 353[1377]); "cross-regional integrations of neural activity" involving the brainstem reticular formation, the thalamus, and somatosensory and cingulate cortices (A. Damasio. 1999. *The Feeling of What Happens: Body and Emotion in the Making of Consciousness.* New York: Harcourt Brace); "selectional reentrant activity of groups of neurons in the [thalamocortical] core" (G. M. Edelman. 2006. *Second Nature: Brain Science and Human Knowledge.* New Haven, CT: Yale University Press); "quantum-coherent oscillations within microtubules" (R. Penrose. 1994. *Shadows of the Mind.* Oxford: Oxford University Press); "the interactions of specialized, modular components in a distributed neural network" (J. W. Cooney and M. S. Gazzaniga. 2003. "Neurological Disorders and the Structure of Human Consciousness." *Trends Cogn Sci* 7[4]); or some other physical or functional state.

20. To see the impasse more clearly, it might be useful to consider a neuroscientific account of consciousness that proceeds with the usual buoyant disregard for this philosophical terrain. The neuroscientists Gerald Edelman and Giulio Tononi claim that it is the intrinsic "integration," or unity, of consciousness that provides the best clue to its physical character. In their view, consciousness is a "unified neural process" born of "ongoing, recursive, highly parallel signaling within and among brain areas." (Gerald M. Edelman

and Giulio Tononi. 2002. *A Universe of Consciousness: How Matter Becomes Imagination.* New York: Basic Books; G. Tononi and G. M. Edelman. 1998. "Consciousness and Complexity." *Science* 282[5395].) Accounting for why the highly synchronous activities of generalized seizures and slow-wave sleep do not suffice for consciousness, the authors provide another criterion: The "repertoire of differentiated neural states" must be large rather than small. Consciousness, therefore, is intrinsically "integrated" and "differentiated." The fact that over a long enough time scale, the entire brain may be said to display such characteristics demands another caveat—because the entire brain cannot be the locus of consciousness. Thus, the authors declare that such integration and differentiation must occur within a window of a few hundred milliseconds. These criteria together constitute their "dynamic core hypothesis."

Tononi and Edelman have done some fascinating neuroscience, but their research demonstrates how forlorn any empirical results seem when hurled against the mystery of consciousness. The problem is that such work does nothing to render the emergence of consciousness comprehensible. While Tononi and Edelman are probably aware of this fact, they nevertheless announce, arms akimbo, that "a scientific explanation of consciousness is becoming increasingly feasible." (G. Tononi and G. M. Edelman. 1998. p. 1850.)

Why would the difference between consciousness and unconsciousness be a matter of "a distributed neural process that is both highly integrated and highly differentiated"? And why should the time course of such integration be a few hundred milliseconds? What if it were a few hundred years? What if distributed geological processes gave rise to consciousness? Let's just say, for the sake of argument, that they do. This would not explain how consciousness emerges. It would be nothing short of a miracle if mere integration and differentiation among processes in the earth sufficed to make the planet conscious. Is the linkage between neural synchrony and consciousness any more intelligible? No—apart from the fact that we already know that we are conscious.

Consider some other possibilities for emergence: Let us say that there is something that it is like to be a coral reef battered by waves of precisely 0.5 hertz; there is something that it is like to be a 150-mile-per-hour wind gust laying waste to a trailer park (but only if the trailers are made entirely of aluminum); there is something that it is like to be the sum total of New Year's resolutions left unfulfilled. How could such diverse "brains" possibly give rise to consciousness? We have no idea. And yet, if we stipulate that they do, their powers are no less comprehensible than those of the brains we have in our heads. But they are not comprehensible at all, of course—and that is the problem of consciousness.

21. Cited in C. Sagan. 1995. *The Demon-Haunted World: Science as a Candle in the Dark*. New York: Random House, p. 272.

22. This distinction was obvious to many thinkers even before vitalism was discredited. C. D. Broad (1925) summed it up with admirable precision:

> The one and only kind of evidence that we ever have for believing that a thing is alive is that it behaves in certain characteristic ways. E.g., it moves spontaneously, eats, drinks, digests, grows, reproduces, and so on. Now all these are just actions of one body on other bodies. There seems to be no reason whatever to suppose that "being alive" means any more than exhibiting these various forms of bodily behaviour. . . . But the position about consciousness, certainly seems to be very different. It is perfectly true that an essential part of our evidence for believing that anything but ourselves has a mind and is having such and such experiences is that it performs certain characteristic bodily movements in certain situations. . . . But it is plain that our observation of the behavior of external bodies is not our only or our primary ground for asserting the existence of minds and mental processes. And it seems to me equally plain that by "having a mind" we do not mean simply "behaving in such and such ways." (Cited in A. Beckermann. 2000. "The Reductive Explainability of Phenomenal Consciousness." In *Neural Correlates of Consciousness: Empirical and Conceptual Questions*, ed. T. Metzinger. Cambridge, MA: MIT Press, p. 49).

23. Another way of stating the matter is that if, as all physicalists believe, there is a necessary connection between the physical and the phenomenal, we would not expect to see evidence for it—apart from the reliability of correlation itself. If we are told that phenomenal state X is really brain state Y, we must ask, "By virtue of what is this identity true?" The answer must be that one cannot find X without Y or Y without X. But this disgorges two further facts: Such an identity can be established only by virtue of empirical correlations, and the phenomenal term is in no way subordinate, with respect to defining what a state is, to its physical correlate. As Donald Davidson said, "If some mental events are physical events, this makes them no more physical than mental. Identity is a symmetrical relation." (D. Davidson. 1987. "Knowing One's Own Mind." *Proceedings and Addresses of the American Philosophical Association* 61.) Brain state Y is identifiable as phenomenal state X only by virtue of its *X-ness.*

The problem is further complicated by the fact that the neural correlates of conscious states seem liable to be a far more heterogeneous class of events than I have indicated. This raises the issue of *multiple realizability*: the possibility that different physical states may be capable of producing consciousness. Finding one such state (or class of states) to be reliably correlated with consciousness would not necessarily reveal anything about the possibilities of consciousness in other physical systems. Multiple realizability is especially problematic for any theory that seeks to reduce consciousness to a specific type of brain state (i.e., any "type-type identity" theory of consciousness). In neuroanatomical terms, we know that a limited form of multiple realizability must be true, because different species of birds and mammals perform many of the same cognitive operations with importantly different neuronal architectures. Of course, it is conceivable that only human beings are conscious, or that consciousness may be instantiated in precisely the same neural circuits in dissimilar brains—but both these propositions strike me as extremely doubtful.

Whatever one's ontological bias, the meaningfulness of correlation depends on the belief that a causal linkage (if not identity) exists between physical states and subjective experience. And yet, correlation is itself the only basis for establishing this linkage. This is not merely a case of Humean angst with respect to causation: We are blind to the physical causes of phenomenal events to a much greater degree than we are to the physical causes of physical events. In fact, Hume's skepticism about our knowledge of causation has not aged very well. Even rats appear to intuit causal connections beyond mere correlations. One can also argue that our ability to pick out individual events in a temporal sequence, or to group events into categories, is the product of causal reasoning. (See M. R. Waldmann, Y. Hagmayer, and A. P. Blaisdell. 2006. "Beyond the Information Given: Causal Models in Learning and Reasoning." *Current Directions in Psychological Science* 15[6]; M. J. Buehner and P. W. Cheng. 2005. "Causal Learning." In *The Cambridge Handbook of Thinking and Reasoning*, ed. K. J. Holyoak and R. G. Morrison. New York: Cambridge University Press.) When I break a pencil, the force applied to it by my hands and its subsequent breaking are correlated, but not merely so. There is much to be said about the microstructure of pencils that makes their brittleness, and hence the observed correlation, intelligible. With consciousness, however, the link appears to be brute. As Chalmers and others have noted, the question remains: Why should such events in the brain be *experienced* at all? (D. J. Chalmers. 1995. "The Puzzle of Conscious Experience." *Sci Am* 273[6]; Chalmers, *The Conscious Mind*; D. J. Chalmers. 1997. "Moving Forward on the Problem of Consciousness." *Journal of Consciousness Studies* 4[1].) But this does not stop neuroscientists and philosophers from trying to simply ram through explanatory analogies that don't quite fit.

24. W. Singer. 1999. "Neuronal Synchrony: A Versatile Code for the Definition of Relations?" *Neuron* 24(1).

25. For doubts on this point, see M. N. Shadlen and J. A. Movshon. 1999. "Synchrony Unbound: A Critical Evaluation of the Temporal Binding Hypothesis." *Neuron* 24(1).

26. Prinz also observes that binding and consciousness are fully dissociable. J. Prinz. 2001. "Functionalism, Dualism and Consciousness." In *Philosophy and the Neurosciences*, ed. W. Bechtel et al. Oxford: Blackwell.

27. A. Polonsky et al. 2000. "Neuronal Activity in Human Primary Visual Cortex Correlates with Perception During Binocular Rivalry." *Nat Neurosci* 3(11); G. Rees, G. Kreiman, and C. Koch. 2002. "Neural Correlates of Consciousness in Humans." *Nat Rev Neurosci* 3(4); F. Crick and C. Koch. 1998. "Consciousness and Neuroscience." *Cerebral Cortex* 8; F. Crick and C. Koch. 1999. "The Unconscious Humunculus." In *The Neural Correlates of Consciousness*, ed. T. Metzinger. Cambridge, MA: MIT Press; F. Crick and C. Koch. 2003. "A Framework for Consciousness." *Nat Neurosci* 6(2); J. D. Haynes. 2009. "Decoding Visual Consciousness from Human Brain Signals." *Trends Cogn Sci* 13(5).

28. Statistics available at www.gallup.com.

29. G. M. Bogen and J. E. Bogen. 1986. "On the Relationship of Cerebral Duality to Creativity." *Bull Clin Neurosci* 51.

30. J. E. Bogen, R. W. Sperry, and P. J. Vogel. 1969. "Addendum: Commissural Section and Propagation of Seizures." In *Basic Mechanisms of the Epilepsies*, ed. Jasper et al. Boston: Little, Brown; E. Zaidel, M. Iacoboni, D. Zaidel, and J. E. Bogen. 2003. "The Callosal Syndromes." In *Clinical Neuropsychology.* Oxford: Oxford University Press; E. Zaidel, D. W. Zaidel, and J. Bogen. Undated. "The Split Brain." www.its.caltech.edu/~jbogen/text/ref130.htm.

31. M. S. Gazzaniga, J. E. Bogen, and R. W. Sperry. 1965. "Observations on Visual Perception after Disconnexion of the Cerebral Hemispheres in Man." *Brain* 88(2); R. W. Sperry. 1961. "Cerebral Organization and Behavior: The Split Brain Behaves in Many Respects Like Two Separate Brains, Providing New Research Possibilities." *Science* 133(3466); R. W. Sperry. 1968. "Hemisphere Deconnection and Unity in Conscious Awareness." *Am Psychol* 23(10); R. W. Sperry, E. Zaidel, and D. Zaidel. 1979. "Self Recognition and Social Awareness in the Deconnected Minor Hemisphere." *Neuropsychologia* 17(2).

32. R. Sperry. 1982. "Some Effects of Disconnecting the Cerebral Hemispheres. Nobel Lecture, 8 December 1981." *Biosci Rep* 2(5).

33. R. E. Myers and R. W. Sperry. 1958. "Interhemispheric Communication through the Corpus Callosum: Mnemonic Carry-over between the Hemispheres." *AMA Arch Neurol Psychiatry* 80(3); Sperry, "Cerebral Organization and Behavior."

34. M. S. Gazzaniga, J. E. Bogen, and R. W. Sperry. 1962. "Some Functional Effects of Sectioning the Cerebral Commissures in Man." *Proc Natl Acad Sci USA* 48.

35. Zaidel et al., "The Callosal Syndromes"; Zaidel, Zaidel, and Bogen, "The Split Brain."

36. K. R. Popper and J. C. Eccles. [1977] 1993. *The Self and Its Brain.* London: Routledge.

37. See C. E. Marks. 1980. *Commissurotomy, Consciousness, and the Unity of Mind.* Montgomery, VT: Bradford Books; J. E. Bogen. 1997. "Does Cognition in the Disconnected Right Hemisphere Require Right Hemisphere Possession of Language?" *Brain Lang* 57(1).

38. T. Nørretranders. 1998. *The User Illusion: Cutting Consciousness Down to Size.* New York: Viking.

39. V. Mark. 1996. "Conflicting Communicative Behavior in a Split-Brain Patient: Support for Dual Consciousness." In *Toward a Science of Consciousness: The First Tucson Discussions and Debates*, ed. S. Hameroff, A. W. Kaszniak, and A. C. Scott. Cambridge, MA: MIT Press.

40. Sperry, "Some Effects of Disconnecting the Cerebral Hemispheres."

41. J. J. Schmitt, W. Hartje, and K. Willmes. 1997. "Hemispheric Asymmetry in the Recognition of Emotional Attitude Conveyed by Facial Expression, Prosody and Propositional Speech." *Cortex* 33(1).

42. J. Blair, D. R. Mitchell, and K. Blair. 2005. *The Psychopath: Emotion and the Brain.* Malden, MA: Blackwell.

43. Most of the studies involved have relied on the Wada test, in which sodium amobarbital is injected into the left or right carotid artery, temporarily anesthetizing the hemisphere on the same side. Researchers have found that anesthesia of the left hemisphere is often associated with depression, whereas anesthesia of the right can lead to euphoria. The literature on stroke has tended to support this lateralization of mood, correlating left-hemisphere strokes with depression, but some studies have put this interpretation in question. See A. J. Carson et al. 2000. "Depression after Stroke and Lesion Location: A Systematic Review." *Lancet* 356(9224); D. W.

Desmond et al. 2003. "Ischemic Stroke and Depression." *J Int Neuropsychol Soc* 9(3).

Research on normal brains has shown that negative emotions such as disgust, anxiety, and sadness tend to be associated with right-hemisphere activity, whereas happiness is associated with activity on the left. However, it might be better to think about these emotional asymmetries in terms of "approach" and "withdrawal," because anger, a classically negative emotion, is also correlated with activity in the left hemisphere. (E. Harmon-Jones, P. A. Gable, and C. K. Peterson. 2010. "The Role of Asymmetric Frontal Cortical Activity in Emotion-Related Phenomena: A Review and Update." *Biol Psychol* 84[3]: 451–62.)

The lateralized presentation of films suggests that the right hemisphere is more responsive to their emotional content, particularly if it is negative. (W. Wittling and R. Roschmann. 1993. "Emotion-Related Hemisphere Asymmetry: Subjective Emotional Responses to Laterally Presented Films." *Cortex* 29[3].) It is also faster than the left to recognize the emotional charge of individual words (*stupid, beautiful*), and in people suffering from depression, it shows a performance bias for negative words. (R. A. Atchley, S. S. Ilardi, and A. Enloe. 2003. "Hemispheric Asymmetry in the Processing of Emotional Content in Word Meanings: The Effect of Current and Past Depression." *Brain Lang* 84[1].) The fact that primates lack direct connections between the right and left amygdalae (regions in the temporal lobes that are especially sensitive to emotionally significant events) suggests an anatomical basis for lateral differences in mood. (R. W. Doty. 1998. "The Five Mysteries of the Mind, and Their Consequences." *Neuropsychologia* 36[10].) The role of the amygdala in our emotional lives, particularly with respect to fear, is very well established. (Joseph E. LeDoux. 2002. *Synaptic Self: How Our Brains Become Who We Are.* New York: Viking.)

44. Popper and Eccles, *The Self and Its Brain.*
45. Zaidel, Zaidel, and Bogen, "The Split Brain."
46. Myers and Sperry, "Interhemispheric Communication through the Corpus Callosum."
47. Bogen, "On the Relationship of Cerebral Duality to Creativity."

48. R. Puccetti. 1981. "The Case for Mental Duality: Evidence from Split-Brain Data and Other Considerations." *Behavioral and Brain Sciences* 4: 93–123.

49. W. James. 1950 [1890]. *The Principles of Psychology* (Vol. I). Dover Publications, p. 251.

50. However, as Dennett points out, it can be difficult (or impossible) to distinguish what was experienced and then forgotten from what was never experienced in the first place. See his insightful discussion of Orwellian versus Stalinesque processes in cognition: D. C. Dennett. 1991. *Consciousness Explained.* Boston: Little, Brown, pp. 116–25. This ambiguity is largely attributable to the fact that the contents of consciousness must be integrated over time—around 100–200ms. (Crick and Koch, "A Framework for Consciousness.") This period of integration allows the sensation of touching an object and the associated visual perception of doing so—which objectively arrive at the cortex at different times—to be experienced as though they were simultaneous. Consciousness, therefore, is dependent upon what is generally known as "working memory."

   Many researchers have drawn this connection: J. M. Fuster. 2003. *Cortex and Mind: Unifying Cognition.* Oxford: Oxford University Press; P. Thagard and B. Aubie. 2008. "Emotional Consciousness: A Neural Model of How Cognitive Appraisal and Somatic Perception Interact to Produce Qualitative Experience." *Conscious Cogn* 17(3); B. J. Baars and S. Franklin. 2003. "How Conscious Experience and Working Memory Interact." *Trends Cogn Sci* 7(4). And the principle is somewhat more loosely captured by Edelman's notion of consciousness as "the remembered present": G. M. Edelman. 1989. *The Remembered Present: A Biological Theory of Consciousness.* New York: Basic Books.

51. L. Naccache and S. Dehaene. 2001. "Unconscious Semantic Priming Extends to Novel Unseen Stimuli." *Cognition* 80(3). Though several studies indicate that the priming stimulus must at least be attended to: M. Finkbeiner and K. I. Forster. 2008. "Attention, Intention and Domain-Specific Processing." *Trends Cogn Sci* 12(2).

52. M. Pessiglione et al. 2007. "How the Brain Translates Money into Force: A Neuroimaging Study of Subliminal Motivation." *Science* 316(5826).

53. P. J. Whalen et al. 1998. "Masked Presentations of Emotional Facial Expressions Modulate Amygdala Activity without Explicit Knowledge." *J Neurosci* 18(1); L. Naccache et al. 2005. "A Direct Intracranial Record of Emotions Evoked by Subliminal Words." *Proc Natl Acad Sci USA* 102(21).

54. D. L. Schacter. 1987. "Implicit Expressions of Memory in Organic Amnesia: Learning of New Facts and Associations." *Hum Neurobiol* 6(2).

55. L. R. Squire and R. McKee. 1992. "Influence of Prior Events on Cognitive Judgments in Amnesia." *J Exp Psychol Learn Mem Cogn* 18(1).

56. M. M. Keane et al. 1997. "Intact and Impaired Conceptual Memory Processes in Amnesia." *Neuropsychology* 11(1).

57. Other phenomena distinguish consciousness from our unconscious mental lives. For instance, certain people suffer a condition called "blindsight," which results from damage to their primary visual cortex. As a matter of conscious experience, they are blind (or blind within a region of their visual field), and yet they can accurately describe the visual properties of objects. They experience this as purely a matter of guessing—after all, they have no experience of seeing—but they manage to "guess" with near perfect accuracy. They are seeing without knowing that they are seeing. (L. Weiskrantz. 1996. "Blindsight Revisited." *Curr Opin Neurobiol* 6[2]; L. Weiskrantz. 2002. "Prime-Sight and Blindsight." *Conscious Cogn* 11[4]; L. Weiskrantz. 2008. "Is Blindsight Just Degraded Normal Vision?" *Exp Brain Res* 192[3].)

58. S. Harris. 2004. *The End of Faith*, New York: Norton, pp. 173–75, 275–77; S. Harris. 2010. *The Moral Landscape*. New York: Free Press.

Chapter 3: The Riddle of the Self

1. Nanamoli. 1995. *Majjhima Nikaya: Culamalunkya Sutta*. Boston: Wisdom Publications. p. 534.

2. It is occasionally said that spiritual practice leads to the experience of "bliss" and that consciousness itself is inherently blissful. How are we to understand this? The term *bliss* does not get much use in West-

ern discourse—and should one ever have occasion to utter it, it will place one's listeners immediately on their guard. Even with reference to sex the word smacks of grandiosity, as though one were asserting something unique about one's capacity for pleasure. A contemplative who speaks of "spiritual bliss" seems to be making an unusual claim to pleasure, to be luxuriating in obscure stirrings of his nervous system, and this does not engender respect anywhere but among the similarly engorged. One who would spend hours each day absorbed in the bliss of meditation seems rather like a heroin addict or an onanist who has transcended the use of his hands. To find a fount of bliss somewhere within one's nervous system is simply undignified.

But an empirical claim here stands to be tested. The claim is that consciousness, prior to self-representation, is intrinsically "blissful." It is not a matter of a gross thrill or a constant feeling of joy, but there is a feeling tone to consciousness, and once realized, it can be felt to permeate every aspect of experience. This is how it can be said in the teachings of Buddhist and Hindu Tantra that "desire arises as bliss," for indeed it can—if desire is recognized as a mere inflection of consciousness. Of course, if desire is not recognized but merely felt, then it arises as a problem to be solved by the acquisition of its object. It is in this sense that desire is generally described as an obstacle to meditation.

3. D. Parfit. 1984. *Reasons and Persons.* Oxford: Clarendon Press, pp. 279–80.

4. The Scottish philosopher David Hume, for instance, saw the problem quite clearly:

> There are some philosophers who imagine we are every moment intimately conscious of what we call our self; that we feel its existence and its continuance in existence; and are certain, beyond the evidence of a demonstration, both of its perfect identity and simplicity. . . . Unluckily all these positive assertions are contrary to that very experience which is pleaded for them; nor have we any idea of self, after the manner it is here explained. For, from what impression could this idea be derived? . . . If any impression gives rise to the idea of self, that impression must continue invariably the same, through the whole course of our lives; since self is supposed

to exist after that manner. But there is no impression constant and invariable. Pain and pleasure, grief and joy, passions and sensations succeed each other, and never all exist at the same time. It cannot therefore be from any of these impressions, or from any other, that the idea of self is derived; and consequently there is no such idea. . . . For my part, when I enter most intimately into what I call myself, I always stumble on some particular perception or other, of heat or cold, light or shade, love or hatred, pain or pleasure. I never can catch myself at any time without a perception, and never can observe any thing but the perception. When my perceptions are removed for any time, as by sound sleep, so long am I insensible of myself, and may truly be said not to exist. And were all my perceptions removed by death, and could I neither think, nor feel, nor see, nor love, nor hate, after the dissolution of my body, I should be entirely annihilated, nor do I conceive what is further requisite to make me a perfect nonentity. If any one, upon serious and unprejudiced reflection, thinks he has a different notion of himself, I must confess I can reason no longer with him. All I can allow him is, that he may be in the right as well as I, and that we are essentially different in this particular. He may, perhaps, perceive something simple and continued, which he calls himself; though I am certain there is no such principle in me. (D. Hume. *Treatise of Human Nature*, Book 1, Section 6.)

5. R. A. Emmons and M. E. McCullough. 2003. "Counting Blessings Versus Burdens: An Experimental Investigation of Gratitude and Subjective Well-Being in Daily Life." *Journal of Personality and Social Psychology* 84 (2): 377–89.

6. Needless to say, I packed at first light and found a new hotel. Upon checking in, I described my morning's ordeal to the man at the front desk, expecting him to be amused to hear how bad things were under the roof of one of his competitors: *The rat was not only in my room, it was in the bed, under the covers.* He remained silent for a long moment, looking vaguely bored. I began to wonder if I had misjudged his English. "We have rats too," he said, as he handed me my key.

7. Tulku Urgyen Rinpoche. 2004. *Rainbow Painting*. Hong Kong: Rangjung Yeshe Publications, p. 53.

8. M. Botvinick and J. Cohen. 1998. "Rubber Hands 'Feel' Touch That Eyes See." *Nature* 391(6669): 756.

9. V. I. Petkova and H. H. Ehrsson. 2008. "If I Were You: Perceptual Illusion of Body Swapping." *PLoS ONE* 3(12): e3832.

10. *Thought insertion* is the sense that thoughts are being placed in one's mind by others. The *delusion of control* is the belief that one's actions and impulses are being controlled by an external force (such as a television or alien beings).

11. Charles Darwin seems to have been the first to perform a test of this sort, by merely exposing two orangutans to a mirror. The modern version of this test was brought into prominence by the work of Gordon Gallup in the 1970s.

12. For a related argument, see A. Morin. 2002. "Right Hemispheric Self-Awareness: A Critical Assessment." *Conscious Cogn* 11(3): 396–401.

13. N. Breen, D. Caine, and M. Coltheart. 2001. "Mirrored-Self Misidentification: Two Cases of Focal Onset Dementia." *Neurocase* 7(3): 239–54.

14. D. Premack and G. Woodruff. 1978. "Chimpanzee Problem-Solving: A Test for Comprehension." *Science* 202(4367): 532–35; C. D. Frith and U. Frith. 2006. "The Neural Basis of Mentalizing." *Neuron* 50(4): 531–34; U. Frith, J. Morton, and A. M. Leslie. 1991. "The Cognitive Basis of a Biological Disorder: Autism." *Trends Neurosci* 14(10): 433–38; S. Baron-Cohen. 1995. *Mindblindness: An Essay on Autism and Theory of Mind*. Cambridge, MA: MIT Press; K. Vogeley et al. 2001. "Mind Reading: Neural Mechanisms of Theory of Mind and Self-Perspective." *Neuroimage* 14(1), Pt. 1; D. C. Dennett. 1987. *The Intentional Stance*. Cambridge, MA: MIT Press.

15. J. Delacour. 1995. "An Introduction to the Biology of Consciousness." *Neuropsychologia* 33(9): 1061–74; E. Goldberg. 2001. *The Executive Brain: Frontal Lobes and the Civilized Mind*. Oxford: Oxford University Press; F. Happe. 2003. "Theory of Mind and the Self." *Ann N Y Acad Sci* 1001: 134–44; M. Iacoboni. 2008. *Mirroring People: The New Science of How We Connect with Others*.

New York: Farrar, Straus and Giroux; M. Merleau-Ponty. 1964. *The Primacy of Perception, and Other Essays on Phenomenological Psychology, the Philosophy of Art, History, and Politics.* Northwestern University Studies in Phenomenology and Existential Philosophy. Evanston, IL: Northwestern University Press; V. S. Ramachandran. "The Neurology of Self-Awareness." Undated. Edge.org; J.-P. Sartre. [1956] 1994. *Being and Nothingness*, trans. H. E. Barnes. New York: Gramercy Books.

16. K. Vogeley et al. 1995. "Mind Reading: Neural Mechanisms of Theory of Mind and Self-Perspective" and P. C. Fletcher et al. 1995. "Other Minds in the Brain: A Functional Imaging Study of 'Theory of Mind' in Story Comprehension." *Cognition* 57(2) use the same story as a stimulus. Saxe and Kanwisher also take the same basic approach: R. Saxe and N. Kanwisher. 2003. "People Thinking about Thinking People: The Role of the Temporo-parietal Junction in 'Theory of Mind.'" *Neuroimage* 19(4).

17. Sartre, *Being and Nothingness*.

18. It seems intuitively obvious that there is a necessary connection between having a sense of self (as opposed to a perfectly nondualistic perception of the world) and the social experience of "self-consciousness." The latter phenomenon appears to be an inflection of the former—in the same way that feeling an object's hardness is just a special case of feeling its solidity. As with so much that interests us about the world, there seems little chance of our proving this connection in a rigorous way. It falls to anyone who would dissociate these concepts to describe a case of self-consciousness that does not entail the experience of selfhood, and an experience of selfhood that does not admit of the possibility of self-consciousness.

19. Ramachandran, "The Neurology of Self-Awareness."

20. J. T. Kaplan and M. Iacoboni. 2006. "Getting a Grip on Other Minds: Mirror Neurons, Intention Understanding, and Cognitive Empathy." *Soc Neurosci* 1(3–4): 175–83; I. Molnar-Szakacs, J. Kaplan, P. M. Greenfield, and M. Iacoboni. 2006. "Observing Complex Action Sequences: The Role of the Fronto-Parietal Mirror Neuron System." *Neuroimage* 33(3): 923–35.

21. Iacoboni, *Mirroring People*, pp. 132–45; M. Iacoboni and

M. Dapretto. 2006. "The Mirror Neuron System and the Consequences of Its Dysfunction." *Nat Rev Neurosci* 7(12): 942–51.

22. M. Dapretto, M. S. Davies, J. H. Pfeifer, A. A. Scott, M. Sigman,
S. Y. Bookheimer, and M. Iacoboni. 2006. "Understanding Emo-
tions in Others: Mirror Neuron Dysfunction in Children with
Autism Spectrum Disorders." *Nat Neurosci* 9(1): 28–30.

23. J. S. Mascaro et al. 2012. "Compassion Meditation Enhances Empathic Accuracy and Related Neural Activity." In *Social Cognitive
and Affective Neuroscience*. September 5. doi:10.1093/scan/nss095.
While findings of this kind are certainly interesting, the jury is
still out on the significance of mirror neurons. And we should not
forget that despite the presence of mirror neurons in their brains,
monkeys lack language and TOM. They also show very little in the
way of empathy.

Chapter 4: Meditation

1. M. A. Killingsworth and D. T. Gilbert. 2010. "A Wandering Mind
Is an Unhappy Mind." *Science* 330: 932.

2. M. E. Raichle et al. 2001. "A Default Mode of Brain Function."
Proc Natl Acad Sci USA 98(2): 676–82.

3. A. D'Argembeau et al. 2008. "Self-Reflection across Time: Cortical Midline Structures Differentiate between Present and Past
Selves." *Soc Cogn Affect Neurosci* 3(3): 244–52; D. A. Gusnard et
al. 2001. "Medial Prefrontal Cortex and Self-Referential Mental
Activity: Relation to a Default Mode of Brain Function." *Proc
Natl Acad Sci USA* 98(7): 4259–64; J. P. Mitchell, C. N. Macrae,
and M. R. Banaji. 2006. "Dissociable Medial Prefrontal Contributions to Judgments of Similar and Dissimilar Others." *Neuron*
50(4): 655–63; J. M. Moran et al. 2006. "Neuroanatomical Evidence for Distinct Cognitive and Affective Components of Self."
J Cogn Neurosci 18(9): 1586–94; G. Northoff et al. 2006. "Self-
Referential Processing in Our Brain: A Meta-Analysis of Imaging
Studies on the Self." *Neuroimage* 31(1): 440–57; F. Schneider et al.
2008. "The Resting Brain and Our Self: Self-Relatedness Modulates Resting State Neural Activity in Cortical Midline Structures."
Neuroscience 157(1): 120–31.

4. K. Vogeley et al. 2004. "Neural Correlates of First-Person Perspective as One Constituent of Human Self-Consciousness." *J Cogn Neurosci* 16(5): 817–27. One study compared Eastern and Western differences in self-representation and found that while both groups showed more midline activity when applying personal adjectives to the self than to another person, Chinese subjects also showed the same effect for judgments about their mothers. The experimenters interpreted this to mean that the Chinese harbor a more collectivist conception of the "self." Y. Zhu et al. 2007. "Neural Basis of Cultural Influence on Self-Representation." *Neuroimage* 34(3): 1310–16.

5. Y. I. Sheline et al. 2009. "The Default Mode Network and Self-Referential Processes in Depression." *Proc Natl Acad Sci USA* 106(6): 1942–47.

6. J. A. Brewer et al. 2011. "Meditation Experience Is Associated with Differences in Default Mode Network Activity and Connectivity." *Proc Natl Acad Sci USA* 108(50): 20254–59; Véronique A. Taylor et al. 2011. "Impact of Mindfulness on the Neural Responses to Emotional Pictures in Experienced and Beginner Meditators." *NeuroImage* 57: 1524–33. Psilocybin reduces activity in these brain areas as well, and to an extraordinary degree: Robin L. Carhart-Harris et al. 2012. "Neural Correlates of the Psychedelic State as Determined by fMRI Studies with Psilocybin." *Proceedings of the National Academy of Sciences,* January 23.

7. E. Luders et al. 2012. "The Unique Brain Anatomy of Meditation Practitioners: Alterations in Cortical Gyrification." *Frontiers in Human Neuroscience* 6:34; P. Vestergaard-Poulsen et al. 2009. "Long-Term Meditation Is Associated with Increased Gray Matter Density in the Brain Stem." *Neuroreport* 20: 170–74; S. W. Lazar et al. 2005. "Meditation Experience Is Associated with Increased Cortical Thickness." *Neuroreport* 16: 1893–97; Eileen Luders et al. 2012. "Global and Regional Alterations of Hippocampal Anatomy in Long-Term Meditation Practitioners." *Human Brain Mapping* 34(12): 3369–75.

8. A. Lutz et al. 2012. "Altered Anterior Insula Activation During Anticipation and Experience of Painful Stimuli in Expert Meditators." *Neuroimage* 64: 538–46.

9. F. Zeidan et al. 2011. "Brain Mechanisms Supporting the Modulation of Pain by Mindfulness Meditation." *Pain* 31: 5540–48.

10. R. J. Davidson and B. S McEwen. 2012. "Social Influences on Neuroplasticity: Stress and Interventions to Promote Well-Being." *Nature Neuroscience* 15(5): 689–95.

11. http://www.news.wisc.edu/22370.

12. C. A. Moyer et al. 2011. "Frontal Electroencephalographic Asymmetry Associated With Positive Emotion Is Produced by Very Brief Meditation Training." *Psychological Science* 22(10): 1277–79.

13. S.-L. Keng, M. J. Smoski, and C. J. Robins. 2011. "Effects of Mindfulness on Psychological Health: A Review of Empirical Studies." *Clinical Psychology Review* 31: 1041–56; B. K. Holzel et al. 2011. "How Does Mindfulness Meditation Work? Proposing Mechanisms of Action from a Conceptual and Neural Perspective." *Perspectives on Psychological Science* 6: 537–59.

14. J. S. Mascaro et al. 2012. "Compassion Meditation Enhances Empathic Accuracy and Related Neural Activity." In *Social Cognitive and Affective Neuroscience* 8(1): 48–55.

15. O. M. Klimecki et al. 1991. "Functional Neural Plasticity and Associated Changes in Positive Affect after Compassion Training." *Cerebral Cortex* 23(7): 1552–61.

16. M. E. Kemeny et al. 2012. "Contemplative/Emotion Training Reduces Negative Emotional Behavior and Promotes Prosocial Responses." *Emotion* 12: 338–50.

17. M. Sayadaw. 1957. *Buddhist Meditation and Its Forty Subjects*, trans. U Pe Thin. Buddha Sasana Council Press; M. Sayadaw. 1983. *Thoughts on the Dhamma*. Kandy, Sri Lanka: Buddhist Publication Society; M. Sayadaw. 1985. *The Progress of Insight*, trans. Nyanaponika Thera. Kandy, Sri Lanka: Buddhist Publication Society.

18. R. Maharshi. 1984. *Talks with Sri Ramana Maharshi*. Tiruvanamallai: Sri Ramanashramam, p. 314.

19. D. Godman, ed. 1985. *Be as You Are: The Teachings of Sri Ramana Maharshi*. New York: Arkana, p. 55.

20. E. Mach. 1914. *The Analysis of Sensations and the Relation of the Physical to the Psychical*. Chicago: Open Court, p. 19.

21. D. R. Hofstadter and D. C. Dennett. 1981. *The Mind's I: Fantasies and Reflections on Self and Soul.* New York: Basic Books, pp. 23–33.
22. Ibid., p. 30.

Chapter 5: Gurus, Death, Drugs, and Other Puzzles

1. *The Gateless Gate* (Japanese: *Mumonkan*). http://www.sacred-texts .com/bud/zen/mumonkan.htm.
2. G. Feuerstein. 2006. *Holy Madness: Spirituality, Crazy-Wise Teachers, and Enlightenment.* Rev. and expanded ed. Prescott, AZ: Hohm Press, p. 108.
3. F. FitzGerald. 1981. *Cities on a Hill.* New York: Touchstone.
4. P. Marin. 1979. "Spiritual Obedience." *Harper's* (February), p. 44.
5. E. Weinberger. 1986. *Works on Paper.* New York: New Directions, p. 31.
6. C. Trungpa. 1987. *Cutting Through Spiritual Materialism.* Boston: Shambhala, pp. 173–74.
7. For instance, see https://www.youtube.com/watch?v=otGQqO 2TYMI.

   Osho was by no means the absolute worst the New Age had to offer. There is no question that he harmed many people in the end—and perhaps in the beginning and middle as well—but he wasn't simply a lunatic or a con artist. Osho struck me as a very insightful man who had much to teach but who grew increasingly intoxicated by the power of his role and then went properly bonkers in it. When you spend your days sniffing nitrous oxide, demanding fellatio at forty-five-minute intervals, making sacred gifts of your fingernail clippings, and shopping for your ninety-fourth Rolls Royce, you might wonder whether you've wandered a step or two off the path to liberation.
8. Harris, *The End of Faith*, pp. 295–96.
9. G. D. Falk. 2009. *Stripping the Gurus.* Toronto: Million Monkeys Press.
10. See, for example, D. Radin. 1997. *The Conscious Universe: The Scientific Truth of Psychic Phenomena.* New York: HarperEdge.
11. E. F. Kelly et al. 2007. *Irreducible Mind: Toward a Psychology for the 21st Century.* New York: Rowman and Littlefield, p. 372.

12. Ibid., p. 374.

13. Ibid., p. 371.

14. Even supposed evidence for rebirth—such as when a person, usually a child, is alleged to recall facts that prove he or she is the reincarnate personality of a deceased person—seems impossible to disentangle from the question of psi.

15. E. Alexander. 2012. *Proof of Heaven: A Neurosurgeon's Journey into the Afterlife*. New York: Simon & Schuster, jacket quote.

16. E. Alexander. 2012. "Heaven Is Real: A Doctor's Experience of the Afterlife." *Newsweek*.

17. A. E. Cavanna et al. 2010. "The Neural Correlates of Impaired Consciousness in Coma and Unresponsive States." *Discov Med* 9(48): 431–38.

18. Alex Tsakiris. 2011. "Neurosurgeon Dr. Eben Alexander's Near-Death Experience Defies Medical Model of Consciousness." Skeptico. November 22. http://www.skeptiko.com/154-neurosurgeon-dr-eben-alexander-near-death-experience/.

19. Terence McKenna. 1992. *Food of the Gods*. New York: Bantam Books, pp. 258–59.

20. The general differences between neurosurgeons and neuroscientists may explain some of Alexander's errors. The distinction in expertise is very easy to see when viewed from the other side: If a neuroscientist were handed a drill and a scalpel and told to operate on a living person's brain, the result would be horrific. From a scientific point of view, Alexander's performance is no prettier. He has surely killed the patient, but the man won't stop drilling. In fact, he may have helped kill *Newsweek*, which announced immediately after his article ran that it would no longer publish a print edition.

21. A wide literature now suggests that MDMA can damage serotonin-producing neurons and decrease levels of serotonin in the brain. There are credible claims, however, that many of these studies used poor controls or dosages in lab animals that were too high to model human use of the drug.

22. Robin L. Carhart-Harris et al. 2011. "Neural Correlates of the Psychedelic State as Determined by fMRI Studies with Psilocybin." *Proc Natl Acad Sci USA*. December 20. http://www.pnas.org/content/early/2012/01/17/1119598109.

23. Terence McKenna is one person I regret not getting to know. Unfortunately, he died from brain cancer in 2000, at the age of fifty-three. His books are well worth reading, but he was, above all, an amazing speaker. It is true that his eloquence often led him to adopt positions that can only be described (charitably) as "wacky," but he was undeniably brilliant and always worth listening to.

24. It is important to note that MDMA doesn't tend to have these properties—and many people would say that it shouldn't be considered a psychedelic at all. The terms *empathogen* and *entactogen* have been used to describe MDMA and other compounds whose effect is primarily emotional and pro-social.

25. I should say, however, that there are psychedelic experiences I have not had that appear to deliver a different message. Some people have experiences that, rather than being states in which the boundaries of the self are dissolved, appear to transport the self (in some form) elsewhere. This phenomenon is very common with the drug DMT, and it can lead its initiates to some startling conclusions about the nature of reality. More than anyone else, Terence McKenna was influential in bringing the phenomenology of DMT into prominence.

   DMT is unique among psychedelics for several reasons. Everyone who has tried it seems to agree that it is the most potent hallucinogen available in terms of its effects. It is also, paradoxically, the shortest-acting. Whereas the effects of LSD can last ten hours, the DMT trance dawns in less than a minute and subsides in ten. One reason for such steep pharmacokinetics seems to be that this compound already exists inside the human brain and is readily metabolized by monoaminoxidase. DMT is in the same chemical class as psilocybin and the neurotransmitter serotonin (but, in addition to having an affinity for 5-HT2A receptors, it has been shown to bind to the sigma-1 receptor and modulate Na^+ channels). Its function in the human body remains unknown. Among the many mysteries and insults presented by DMT, it offers a final mockery of our drug laws: Not only have we criminalized naturally occurring substances such as cannabis, but we have criminalized one of our own neurotransmitters. Many users of DMT report being thrust under its influence into an adjacent reality where they are met by alien beings who appear intent upon sharing information and demonstrating the use of inscrutable technologies. The convergence of hundreds of such reports, many from first-time users of the drug who have not been told what to expect, is certainly interesting. It is also worth noting that these accounts are almost entirely free of religious imagery. One appears far more likely to meet extraterrestrials or elves on DMT than traditional saints or angels. I have not tried DMT and have not had an experience of the sort that its users describe, so I don't know what to make of any of this.

26. Of course, James was reporting his experiences with nitrous oxide, which is an anesthetic. Other anesthetics, such as ketamine hydrochloride and phencyclidine hydrochloride (PCP), have similar effects on mood and cognition at low doses. However, these drugs differ from classic psychedelics in many ways—one being that high doses of the latter do not lead to general anesthesia.

27. W. James. 1958. *The Varieties of Religious Experience*. New York: New American Library. p. 298.

INDEX

Page numbers in *italics* refer to illustrations. An *n* following a page number refers to the note section.

sleep, consciousness and, 61–62
Slouching Towards Bethlehem (Didion), 187–88
Smith, Joseph, 25, 152, 168
social bonding, 114
sodium amobarbital, 223*n*
solitude, 1–2, 13–14
somatoparaphrenia, 107
soul, 91, 116
  as illusion, 23, 62, 83
Sperry, Roger, 72
spiritual authority, 151–71
spiritual development, stages of, 46–47
spirituality, 7, 10, 205, 209*n*
  atheist animosity toward term, 6, 11, 202
  common lack of interest in, 83
  enlightenment in, 49
  epiphany in, 84
  happiness and, 17–18, 226*n*
  and the illusion of the self, 9, 82
  mind and, 46, 62
  religion vs., 6, 8–9, 19–23
  science vs., 7–8
  solitude and, 13
split brain, 62–68, 72–75, 223*n*–24*n*
  and personal identity, 84, 88–89
Sri Aurobindo, 169
Sri Lanka, 26
stimulus-independent thought, 119
stress, brain structure and, 121–22
stroke, 223*n*
subjectivity, 53, 91, 205, 212*n*, 216*n*
subliminal responses, 76–77
sudden realization, 124–25
suffering, 38, 40–42, 137
  meditation as relief for, 171
Sufism, 22

talking to oneself, 93–94, 100–101
telekinesis, 212*n*
telepathy, 170
Tendzin, Ösel, 161
theory of mind (TOM), 110–13
Theosophical Society, 24–26
Theravada Buddhism, 34–35, 124, 125
thermodynamics, second law of, 89
thought insertion, 108, 229

thoughts:
  as distraction from mindfulness, 36–39, 45, 93–104, 130–33, 148
  happiness and, 119
  meditation and, 100, 101–2
  mental state and, 96–97
  self and, 128
  spontaneous appearance in consciousness of, 101
  stimulus-independent, 119
  subsidence of, 127
  suffering vs. calm in, 95–96
Tibetan language, 138
Tiruvanamali, 128
tobacco, 187, 188
Tononi, Giulio, 217*n*–18*n*
transcendence, 7, 209*n*
trekchod, 138
Tsoknyi Rinpoche, 138
Tulku Urgyen Rinpoche, 132, 134–38

unconscious:
  Freud and, 75
  processing in brain, 75–77, 215*n*, 226*n*
U Pandita, Sayadaw, 125–26

Vedanta, *see* Advaita Vedanta
vipassana, 34–35, 136
Vishnu, 19
Visuddhimagga, 125
vitalism, 219*n*
Vivekananda, Swami, 26
volition, 104

Wada test, 223*n*
well-being, human, 48
West:
  Eastern spirituality in, 23–33, 130, 166
  gurus in, 152, 159–63
  self-representation in, 232*n*
Wilkins, Charles, 24
World Parliament of Religions (1893), 26
Wright, Frank Lloyd, 155

yoga, 12, 23, 47, 193
  hatha, 26–27

Zen Buddhism, 138, 153, 157